Forget Photography

Forget Photography

Andrew Dewdney

Goldsmiths Press

First published in 2021 by Goldsmiths Press
Goldsmiths, University of London, New Cross
London SE14 6NW

Printed and bound by Versa Press
Distribution by the MIT Press
Cambridge, Massachusetts, and London, England

A CIP record for this book is available from the British Library

ISBN 978-1-912685-82-0 (hbk)
ISBN 978-1-912685-81-3 (ebk)

www.gold.ac.uk/goldsmiths-press

Goldsmiths
UNIVERSITY OF LONDON

For Oscar, Thea, Astrid and Nate

Contents

Figures

Acknowledgements

This book would not have been possible, nor conceived, without having a job at London South Bank University (LSBU) for the last twenty years, a highly supportive and convivial institution at the Elephant and Castle, retaining much of its former polytechnic character and ethos. For the first seven years I was employed in managing a growing portfolio of undergraduate digital media courses in what was a second wave of such expansion in higher education in the UK, which afforded me no time for sustained scholarship. During 2005, by extending my working day and through a creative collaboration with Peter Ride, we were able to produce *The New Media Handbook* for Routledge in 2006, which reignited my interest in academic study. There is a much longer history to my interest in what was once the medium of photography and how it could be mobilized, dating back to my time at the Cockpit Arts Workshop, The Watershed Media Centre in Bristol and at Newport School of Art and Design. Over my years at LSBU I have had the support and encouragement of many serious scholars in social science and media sociology, in particular Jeffrey Weeks and Janet Holland. In its most recent incarnation, the School of Arts and Creative Industries, under the stewardship of Janet Jones, has provided me with a safe haven in which to pursue more focused research. Phil Hammond has been a stalwart supporter of humanities research in an otherwise technocratic institution and I would like to thank him for his steady hand. More recently, Geoff Cox and Elena Marchevska, colleagues whose opinions I value and judgements I trust, have provided me with strong support. I would also like to acknowledge my colleagues Daniel Alexander, Adam Brown and Simon Terrill for serious and playful discussions held in offices, the pub and the Art Café, speculating on the future and fate of photography education. In 2007, I was fortunate to be the recipient of a successful Arts and Humanities Research Council research award in collaboration with Tate Britain, entitled *Tate Encounters: Britishness and Visual Culture*, and thus began a long and fruitful intellectual friendship with Victoria Walsh, whose sharp and fearless mind has helped me ever since, including a critical reading of the original book proposal.

In 2012, I cofounded the Centre for the Study of the Network Image (CSNI) at LSBU with Katrina Sluis and Daniel Rubinstein, who shared an office next to mine in Borough Road. For a time we seemed to be in constant discussion and argument about photography education and the state of photography in general and that period remains seminal to this project. Annet Dekker arrived as the first postdoctoral administrator of CSNI and joined us as a co-director. Since then I have continued to work with Annet in joint supervision of our small group of research students and value her expertise, clear thinking and direct way of putting things. Annet also provided a critical commentary on the original book proposal, which helped my thinking. Working in the research environment of CSNI has also had a hand in shaping this book and I would like to acknowledge the contributions of all of my research students to my thinking, in particular Ioanna Zouli, Nicolas Malevé, Gaia Tedone, Lozana Rossenova, Carolyn Defrin, Nicola Baird, Jeannette Ginslov, and more recently Victoria Ivanova, Qian Xiao, Rosie Hermon and Theresa Kneppers, all of whose research topics border upon my own.

Parts of the book have had lecture outings. In particular, Chapter 5 on photography and modernism owes much to joint conference papers written with Victoria Walsh, while the key motif of the book, Chapter 2 on zombie photography, was refined over three lecture outings over a period of five years. I first presented the idea to a group of archivists and conservationists gathered at a Norwegian Arts Council Conference in Oslo in 2014, where the digitisation of photographic collections was under discussion and where I met Nina Lager Vestberg and Anna Dahlgren, who took a lively interest in my thinking. Subsequently I gave a shorter version of the zombie photography lecture at the 2015 Annual Conference of the UK Association of Photographers in Higher Education at Nottingham University. The last outing of zombie photography prior to writing the book was at a research seminar held by the Post-Photographic Research Project, directed by Wolfgang Brückle at the Hochschule Luzern – Design & Kunst in 2019. During the latter stages of writing I was extremely lucky that the Post-Photography Research reading group of Katrina Sluis and Wolfgang Brückle, together with Doris Gassert and Marco de Mutiis, both from FotoMuseum Winterthur, read the first two chapters and gave me encouraging critical feedback at a Zoom meeting.

In 2019, Goldsmiths Press accepted my proposal and I am deeply grateful to Sarah Kember for the risk she took and to the editorial board for endorsing it. Since finishing a first draft, Susan Kelly has been most helpful and responsive as the commissioning editor and I am more than grateful to Dan Shutt for editing the manuscript and tolerating my innumeracy when it comes to dates. Goldsmiths would not have taken it on, I think, if it were not for the supportive critical reviews provided by Ben Burbridge and Katrina Sluis, whose help in seeing the book's flaws has been invaluable. Their commentary has proved a trusted guide in clarifying the arguments over the course of revising the book during the Covid pandemic. I extend a particular debt of gratitude to Katrina Sluis, who has been so very generous in the level of detailed commentary on the entire manuscript and for providing constant encouragement. I would also like to acknowledge the long-term support of Chris Miller and Martin Lister, old friends on whom in the early 'wilderness' days of writing I pressed first drafts and who responded generously in reading, commenting and generally believing in me and the project. I would also like to thank Robin Marriner for taking the time to read the first three chapters and whose email responses have led to further interesting discussions about theoretical writing and art school education. I would also like to thank Stephen Wilson, who I reached out to on Max Dewdney's instigation and who read Chapter 2 on zombie photography, giving me yet another insightful and positive response.

Writing the book has preoccupied my waking hours for over a year and it could not have been sustained without the tolerance, nurture and love of my lifelong partner Catherine, who has maintained and ordered the world around me during this indulgence, keeping me connected to my children Alice and Max, their partners Jeremy and Helen and our four grandchildren in addition to discussing my daily progress over supper. The book is the outcome of a rich network of love, encouragement and critical interest for which I am truly grateful.

Part I

1

Forget Photography

The tradition of all dead generations weighs like a nightmare on the brains of the living. And just as they seem to be occupied with revolutionizing themselves and things, creating something that did not exist before, precisely in such epochs of revolutionary crisis they anxiously conjure up the spirits of the past to their service, borrowing from them names, battle slogans, and costumes in order to present this new scene in world history in time-honored disguise and borrowed language.

Karl Marx, *The Eighteenth Brumaire of Louis Bonaparte*, 1852 (MECW 11), pp. 103–106.

'Theoretical production, like material production,' Baudrillard wrote, 'loses its determinacy and begins to turn around itself, slipping en abyme towards a reality that cannot be found.'

Sylvere Lotringer. 2007. 'Exterminating Angel', introduction to Jean Baudrillard, *Forget Foucault*. Los Angeles, CA: Semiotext(e).

The book argues that if we wish to understand the politics of representation in the post-photographic era, or, more specifically, the image under the conditions of capitalist, computational reproduction, there is a necessary prerequisite, and that is the need to 'forget photography'. The very term photography is a barrier to understanding the altered state of the default visual image. The central paradox this book explores is that at the moment of photography's technical replacement by the screen, algorithm and data flow, photographic cultures proliferate like never before. Photography is everywhere, but not as we have known it; for some time it has been an undead, a zombie, in which the established language, thinking, meanings and values of photography now stand as an obstacle to grasping the new condition. The current mode of image production and circulation turns visual representation on its head and with it is changing how we think about humanness and the world. The image has fled its analogue forms and now haunts the opaque intimacy of the screen and its algorithmic abstractions, creating new questions of how to understand visual meaning, indeed all meaning, in computational and network culture. But, as this book takes up, it also bequeaths us the question of how to regard photography's afterlife. Forgetting photography is not a simple injunction, made at the outset in order to move on to accounts of photography's replacement by the computational networked image. True, the computational networked image is now the default of reproduction and the visual in culture and therefore is the new locus for understanding the power and agency of images. However, the afterlife of photography, residual as it might technically be, also maintains a powerful representational hold on culture and upon reality, which it is important to understand in relation to the new conditions. It is not a question of one medium superseding another, in a teleological version of technical development, nor even finally a question of the remediation of the medium of photography by computing. Both exist in the same temporal space and the effort is to understand the conditions that maintain their state and the relations between them.

Forgetting photography is a strategy to reveal the redundant contemporaneity[1] of the photographic constellation and the cultural immobility of its epicentre. Forgetting photography attempts to put photography into historical perspective and to liberate the image from these historic shackles, forged by art history and photographic theory. More importantly, perhaps, forgetting photography also entails rejecting the frame of reality it prescribed and delineated and in doing so opens up other relationships between bodies, times, events, materials, memory, representation and the image. Forgetting photography attempts to develop a systematic method for revealing the limits and prescriptions of thinking with photography, which no amount of revisionism of post-photographic theory can get beyond. The world urgently needs to unthink photography[2] and go beyond it in order to understand the present constitution of the image as well as the reality and world it has shown and continues to show. Forgetting photography will require a different way of organising knowledge about the visual in culture involving crossing different knowledges of visual culture, science, technologies and mediums. It will also involve thinking differently about routine and creative labour and its knowledge practices within the institutions and organisation of visual reproduction and will therefore inescapably entail politics.

The need to forget photography is now overdue and the need to start somewhere else is urgent if we are to understand current ways of seeing and the reality they conjure up. Forgetting photography can be understood as looking back on photography and laying to rest its place in the culture of twentieth-century industrial capitalism. It can also be understood as the making of a new clearing from which we can look at the present and the interconnections between finance capitalism, proxy wars fought for resources, ecological damage and the new default structures and apparatuses of knowledge and communication with which we think and act. The book is a long argument about forgetting photography and a series of essays about how and where to start building a practically useful and politically engaged different account of the part played by technologies in reproducing reality. The choice between continuing to resuscitate photography in various post-embraces and forgetting it is stark. Forgetting is not easy, not least because it is tied to remembering, but the starting point for this enquiry is an insistence that a decisive break with photography's mode of being, thinking and language is required. Forgetting photography is the hypothesis upon which the book is based and forms the methodological approach in setting out not only why we need to forget photography but in practice how it might be done and what new vistas and approaches it affords. Forgetting photography as we will see has a methodological logic which is polemical, transdisciplinary and transactional in pursuing a problem across intellectual fields and institutional settings and can become a practical way of thinking and doing things. Forgetting photography ultimately seeks to align a knowledge of the image in culture with all progressive struggles for emancipation.

The title of this book is as serious as it is playful and is a direct reference to Jean Baudrillard's (in)famous essay *Forget Foucault* (2007) originally published in 1977, in which Baudrillard declares Foucault's writing to be 'too perfect' in giving an account of what it proposes. Baudrillard argues that as a discourse of power, no longer based on a despotic or catastrophic architecture, Foucault's writing is a seamless, meticulous unfolding of a narrative without origin, in which power 'seeps through the whole porous networks of the social, the mental and of bodies infinitesimally modulating the technologies of power' (2007, p. 29). In short, Baudrillard sees Foucault's discourse as a mirror of the powers it describes and cannot be taken as a discourse of truth, but a mythic discourse, which has no illusions about the effect of the truth it produces. The polemic advanced over the course of this book is that photography, in its analogic perfection, like Foucault's writing, has been 'too perfect' and, in the same manner as Foucault's discourse of power, photography reinforces objectivity and ensures the reality principle. Foucault's too-perfect discourse maintains the principle of objective reality in the same manner as photography, through its exactitude, whereas Baudrillard argued that both power and reality have been disseminated and dissolved by simulation, creating a state of hyper-reality.

The book takes on a number of practical and personal challenges of different scales in order to articulate a different way of thinking about photography, the image and the visual world in networked culture. In making this attempt the biggest challenge and indeed the goal has been to overcome the stubborn, familiar and ingrained lexicon of photography and its visual taxonomy. This is an important task to attempt, not because photography is at stake, but because, more importantly, reality is at stake. This is the existential and affective reality of all human beings on the planet, who make their own lives, but not in circumstances of their own choosing. Such circumstances are given, transmitted from the past and present in the material organisation of the ceaseless forces of accumulation and labour upon which national states, their laws, institutions, military and civil societies have been erected and continue to be maintained. Photography takes part in these arrangements and has prescribed and shaped a representational reality of the twentieth century, a reality that can and needs to be questioned. More to the point is that the image of reality bequeathed by twentieth-century photography was not only ideological and cruel, but no longer matches the conditions of representation, and conceals the conditions of the twenty-first century. The photographic image remains the cultural default for reality, even though the systemic default of the visual image has moved to a nonrepresentational system. The consequence is that the subjective, social and scientific reality that photography encodes seriously occludes the emergent reality of computation and misses the greater present chaos of reality. This is the reality of the interrelationships between human and other-than-human things, what Bruno Latour and others have called hybrids and in other ways what Jean

Baudrillard has called simulation and hyper-reality.[3] It is also the reality of social democratic nation states colluding with capitalism in forms of systemic global oppression in which broadcast media has little or no insight into its own complicity. It is, finally, a reality of the paradoxical present, which, as will be seen, shapes the argument of forgetting photography.

Recalibrating the Visual

One of the main problems the book examines is that photography, as it is still known, takes part in reproducing the reality of the everyday life world whilst appearing to stand apart as a representation in different registers of that reality. In addition, on a common-sense view the photographic image is taken as synonymous with the spontaneous bodily reflex of sight, as natural seeing, and yet the photograph is far from natural; it is overwhelmingly a graphic artifice, a two-dimensional pictorial code of symbolic communication. Photography as part of the reproduction of ways of seeing has, over the course of the twentieth century, become naturalised in the everyday life world as a transparent window on reality, as well as being adopted as a scientific measure and an aesthetic expression. Photography also reproduces itself unknowingly as the unity which is photography. Beyond the practices of everyday life, the production of social and scientific knowledge related to reproduction is formalised at a tertiary level of commercial, state and educational research, in which photography is also involved. Disentangling ideas about photography, ways of seeing and the visual in culture is one of the tasks set out in what follows. The visual in Western culture contains a paradox in which human seeing is both an evolutionary property of the eye and brain as well as something humans collectively construct. The human infant's adjustment to seeing the world is therefore biological and cultural. Seeing is both prewired and learnt. Reality is overwhelmingly remade through routine habits of thought and action within the given conditions of the everyday life world. Jacques Rancière's idea of 'the distribution of the sensible'[4] is a more abstract and political way of putting this (2004, p. 12). The ways in which photography is practised and thought about are also a routine and a habit. The immediate human life world is shaped and constrained by complex interactions of global geo-political and bio-political forces and complexity is what we need to contend with. Photography, as it continues to be understood, takes part in these larger systems of relations of power. Individuals creatively struggle to make sense of their life world and to realise and channel agency in which common ways of seeing are involved; to change the world is to see differently. The formalisation of knowledge produced through disciplinary institutions is selectively distributed in culture through commercial application, compulsory education and broadcast and online media. In the case of photography its knowledge domain is constituted and distributed primarily through practice and education. The relationship between the everyday and formal knowledge transmission about the changes taking place in visual culture needs questioning and challenging, not only from the position of peer review but from the politics of everyday life, because it is an alliance which extends the afterlife of photography.

If photography is no longer photography, then what is it we are doing on our smart phones, cameras and computers, uploading, scrolling, swiping, saving, sharing and printing? Image making has become inherent to life itself. If this is not photography, then what is it? We will see that a number of new terms have come to the aid of the photographic image to help describe its current state, such as the technical, unfettered, fluid, soft, operational, machinic, non-human and, the preferred term here, the networked image. These adjectives are reached for to describe what has been called expanded photography on the one hand, and the computational or algorithmic image on the other. Such linguistic qualifiers of photography betoken the central fact that whatever it has been has changed and that such a situation throws us back upon language as the means by which the visual image has to be rethought. The apprehension of the photographic image cannot easily be divorced from the language used to conceive it and reinforces an understanding that images have a social ontology. The relationship between the image and word and more abstractly the presence of the image in thought through language comes into contention with ideas about the primacy of vision and the acknowledged ascendency of the visual in culture since the second half of the twentieth century, when photography, film and television were seen to be predominant. Popular or mass media have been studied as 'language-like', even up to and including new media (Manovich 2001), in which a science of signs, claimed, with justification, increasing dominion over how meaning operated in visual images. Images were taken as texts to be decoded and thus foregrounded a literary tradition and the centrality of the logos in cultural analysis. Images were also recognised to be part of the psycho-social world, circulating in the unconscious, as much as in media culture, and provided fertile ground for thinking photography in terms of psychoanalysis and semiology (Burgin 1982). Now, the mathematical logic of the computational image seriously disrupts the semiological analysis of the photograph and calls for a new way of understanding the nonrepresentational basis of the image and the new practices it invokes. And yet the photographic image is still, for the most part, received in culture as a representational system of meaning. It is representation, rather than photography, that needs critical attention and the effort to separate the two is part of making the space to consider visual representation in its transmedial and multimodal forms.

The language used to interpret the photographic image and the idea of a photographic language are deeply entwined and rooted in Western philosophical thought and, in the dominant representational mode and its allied ways of seeing, language is deeply embedded and inextricably linked to the visual image. One of the unavoidable problems encountered in attempting to see the historical limits of the language of photography is therefore how to move beyond it – how in effect to account for visuality and representation in common culture in ways which connect the new technical condition of the image with the political traditions of critical analysis and cultural activism. It is still the case that progressive identity politics as well as the environmental activist movement continue to employ the photographic image in representational terms, in contrast to the neo-fascists, who have learnt to 'successfully' exploit network communication as fake news and alternative facts. Recently published academic work points to the fact that the massification of photography is being rethought and recalibrated primarily in relationship to the functions of technology. Such projects inevitably test the boundaries of concepts and language within which newer technologies of vision are accounted for. As yet the language of vision relies on the limited lexicon of the key terms seeing, image, picture and photograph, in themselves complex abstractions, to do a great deal of work in accounting for such major changes, and which are being qualified even more in relationship to technologies of measure and scale (Dvorak and Parikka 2021). As a qualification to the current emphasis upon technologies of vision, W.J.T. Mitchell reminds us that 'The image never appears except in some medium or other, but it is also what transcends media, what can be transferred from one medium to another' (2005, p. 16). The image then is as much a mental process as it is a material arrangement. To paraphrase Mitchell, a photograph is something you can frame and hang on a wall, but the photographic image is what appears in a photograph and, I would add here, both the photograph and the photographic image are constituted as the image of photography. The method of forgetting photography developed here recognises technical and formal distinctions between vision and language, the photograph, word and image, but adopts a pragmatic position in which the relationships between word and image, language and photography are experienced in their encounter and practical uses. The agency of each element is constituted by a network of active associations in which language is always entailed.

The visual in all forms of media operates as a zone of social contact and symbolic exchange between the body, memory and phenomena, or what is lived and felt as the world of external events, and the world as imaged, pictured and represented. Over the course of photography's history, the photographic image has come to occupy a position of unbridled authority in and over the social and scientific real. During the latter part of the twentieth century the veracity of the photograph, founded on its supposed indexical link to external events in time and space, became increasingly contested in practice and theory. In the twenty-first century, concern over the veracity of images has become more muted and mutable through the immersion with screens and graphical user interfaces. This is a moment of the fully fledged post-photographic and as insisted so far there is a pressing need to move beyond it.

The main strategy of the book lies in adopting the view that photography is no longer the active organising mode of the visual image, in effect to see photography as redundant, but still exerting an influence in an afterlife. The afterlife of photography is cast as the fictional figure of the zombie, which is of course playful, but also a serious methodological ploy designed to discuss the limits of photography's academic theorisation as well as its contemporary cultural institutional practices. It is important to stress that the project is not about announcing another 'death' of photography, but rather bringing its afterlife into focus. Photography has been a living dead since the last quarter of the twentieth century. Geoffrey Batchen noted as far back as 1994 that a number of distinct points of death have been plunged into the body of photography, even though he continues to take this as a sign of life in photography (2021), but the argument here is that the totality of these deaths has been for some time photography's after-life. This is an exploration of what this more-than-symbolic death of photography means for knowledge and understanding of the image and as an encouragement to move on. If we want to understand the vexed relationship betweeen image–representation–reality, photography can only assist in this task now as a ruin and an archive, as something whose outlines and conditions only memory can recover.

Photography is a ruined territory populated by archaic knowledge practices bounded by a computational network of relations between images, humans and machines. Such a stark, if not bleak, perspective will, it is hoped, become clear in what follows. Since its inception photography as a transactional information system has been deeply imbricated in military, industrial, commercial, scientific, medical, national, domestic and arts networks of associations, flows and reproductions. The import of such an understanding here, however, is not to repeat this perspective in order to extend the photographic map, nor to inventory the ways in which photography has been and continues to be constituted as a practice field. Rather, it aims at the opposite, to deterritorialise[5] photography, to flee photography in order to force a new view of the image. But in order to do this, to forget photography, it cannot simply be abandoned, even though it has already been lost. There is as yet no outside to photography because it continues to be taken as a default of representation. The way out of photography proposed here is through remembrance, witnessing the trauma of photography's several deaths, from the perspective of its after-life. This is achieved by a trick, by adopting the future present from which contemporary photographic knowledge practices of collection, exhibition and archiving appear as photography's spectral self. It is in institutional knowledge practices that the order of simulation can be identified: the order of representation, modernism, technology, heritage and finally post-photography. By looking at the contemporary state of photographic difference(s) the argument about why we need to forget photography emerges. It is an argument which claims that photographic discourse now conceals more than it reveals about the state of culture, society and the agency of the image. Forgetting photography calls for a more productive discourse in which the hybridity of the networked image, inequality, racism and climate damage stand at the centre of concern.

The approach taken to the task of forgetting photography is to ground speculative and conceptual thinking in the practical affairs and everyday arrangements in which ideas are produced, circulated and received, to see what purposes, functions, connections and advantages are served by the unintended conspiracy to perpetuate the photographic universe. The strategy of the book involves identifying how ideas and practices of what is still taken to be photography move seamlessly across time, the body, everyday life and formalised culture. The overriding conceptual challenge the book takes on is how to intervene productively in the interstices of these related domains and fields of enquiry in order to open up a different way of thinking about the current state of the visual in culture. The argument for forgetting photography is made primarily as an intervention into the (re)production of academic knowledge about photography and this comes with a cost in terms of what has to be taken on, the breadth of material which has to be covered and the technical language required. But writing into academia was not the original hope of the book, which aspired to speak more broadly and accessibly about photographic cultures under the new conditions of computational networks. In completing the book, I realised that making the metaphorical 'new clearing' from which a new theory of the visual image in culture might emerge required much more hacking through the dense undergrowth of academic writing about post-photography than I had anticipated. In attempting to forget photography the book has had to engage just as much with the technical apparatuses of knowledge production as it has with technologies of seeing and, of course, the two are centrally linked by the common condition of commodification and instrumental datafication. Holding the technological apparatuses of both seeing and knowledge together essentially frames the main object of this enquiry as the reproduction of the idea of photography and explains why the lengthy discussions of how to unthink photography's history and theory are necessary.

Organisation of the Book

The book is organised conceptually according to the advice I offer research students, which is to identify a problem in the world, locate its practical and intellectual context, define key questions, work out a method of analysis, apply it to specific situations and report on the results. In trying to follow my own advice and supported by critical comment from colleagues and friends on earlier drafts, the book has finally fallen out in three parts, or, in more dramatic terms, three acts, if not psychological moments, not completely under my control, but of my own making, which, somewhat unevenly, structure the arc of the argument. Emotionally, the book has been motivated by an enduring frustration with successive deformations of the revolutionary spirit of modernity since 1968, then living through and embracing the condition of postmodernity, only to find myself back in a culture of deep conservatism and reaction. The book strives to employ the central analysis and political spirit of Marx and the intellectual movements his work led to and its creative reformulations and adaptations. More specifically, it has been influenced by the ideas of Bruno Latour, Jean Baudrillard, Paul Ricœur and Gilles Lipovetsky, amongst others – a male cast and a strange combination with differing perspectives on the world. Latour was critical of Baudrillard for losing touch with reality, Lipovetsky shared Baudrillard's sense of hypermodernity, whilst they would all share Ricœur's anthropological and phenomenological view that the self is not immediately transparent to itself, but that individuals are agents responsible for their actions. What unites my reading of their different work is an overriding sense that the present is above all paradoxical in terms of both individual life and its multiple and contradictory positions and the general social and global uncertainty about the future of the planet. There is a practical utility in the idea of the paradoxical present, which gives cause for hope in moving away from worn-out certainties and binary conceptions towards a greater understanding of hybridity and hypermodernity. The concept of the paradoxical present is also a productive challenge to the older certainties upon which most socialist and revolutionary movements remain based. Certainly, the argument for forgetting photography is paradoxical since photography is more extant than ever, as well as because forgetting involves remembering. However, making the argument for forgetting photography opens up other productive ways of seeing the current image culture as well as offering different ways of looking at photography's history. The paradoxical position embraced by the argument creates a particular problem of its paradoxical nature. It is argued that photography no longer exists, but at the same time finds signs of its (after)life everywhere. It is argued that the new default of the reproduction of the image requires a new conceptual language, yet continues to use an older set of linguistic terms. The paradoxes abound and, according to the logic of the paradoxical present, can't be otherwise, which makes forgetting photography a provisional and paradoxical enterprise and possibly only a thought experiment. However, the structure of argument is not only a speculation; it is also grounded in analysis of photographic practices. Part II (Chapters 4, 5 and 6) attempts to reveal photographic theory, exhibition and the archive as monuments to photography's past life, erected and maintained through the institutional disciplinary practices of research and curating. Seeing photography as already part of the past opens up space in Part III to attempt to define a situation beyond photography and offer a provisional and imperfect outline of the territory.

In more detail, Chapter 1, 'Forget Photography,' outlines the polemic of why continuing to think with photography masks the objective state of the current mode of capitalist reproduction, and indicates what is at stake and how the currencies of photography are reproduced. One of the main affordances of the argument is that it opens up two new spaces of thought, to consider how new ways of seeing might be elaborated without the historical baggage of photography and how the history of photography can be revisited to reveal its relationship to capitalism and imperialism. In Chapter 2, 'Zombie Photography,' I am indebted to Paul Ricœur for giving me the tools to think of what is involved in the memory of photography, a task which led me via Ariella Azoulay and Jonathan Beller back to Alan Sekula. There has, of course, always been a sporadic socialist history of photography, one which examines photographic practice in relation to the exploitative system of capitalism. However, the strategy of photography's afterlife opens up new and surprising avenues for the remapping of histories. Chapter 3, 'Post-Photography,' lays out the academic development of image studies across art history, cultural and media studies and new media studies from the 1990s. It charts how the object of the idea of photography is both parsed and reconfigured as it travels across fields and sub-fields of knowledge, creating the problem of a lack of cohesion when it comes to understanding the new computational image condition. Post-photography is taken as an inclusive term for all photographic theory since Mitchell coined the term in the 1990s, because essentially the development of the digital image is the backstop to the zombie condition photography finds itself in. For the position of after-photography, post-photography has to be understood as the inclusive term for the continuation of photography. Having made the case for forgetting photography and seeing its affordances, Part II of the book, 'Remembering', constructs three 'case studies' to look in detail at how photography's afterlife masquerades as the default of the contemporary cultural image in the cultural practices of academia and museums. In Chapter 4, 'Philosophy, Technology and Photography', the most difficult to write, I look at the production of knowledge about photography and in particular focus on phenomenological philosophy's continued influence upon thinking about the ontology of photography, arguing that using the abstractions of phenomenology to understand photography elides it with the wider computational apparatus and hence makes no distinction between its material and historical specificity. Another elision, this time between photography and the contemporary, is examined by looking at the exhibition and collection practices of Tate Modern and Tate Britain, arguing that in framing photography as contemporary art they exclude the new contemporary situation of the network image and its Internet ecology. A similar situation unfolds in Chapter 6, 'Photography and Heritage', which examines the expanded collection and photographic galleries at the Victoria and Albert Museum in terms of an equally unresolved view of wanting to collect contemporary digital photography by suturing into a continuous history of photographic art and science. In Part III (Chapters 7, 8 and 9), the address is unequivocally upon the contemporary condition and offers a view of the condition of the network image, a view of the politics of photography and the image and a view of the condition of hypermodern culture and the hybridity of all media images. Chapter 7, 'The Image after Photography', sets out a series of 'transitional steps' in the formation of the image after photography – from analogue inscription to digital dataset, from image apparatuses to social performances and from the discourse of photography to the discourse of computing – landing upon the network image as a provisional definition of the new condition of the image. Chapter 8, 'Hypermodernity,' starts with Lipovetsky's idea of hypermodernity, the paradoxical present and the strategies of chrono-reflexivity as a way of characterising everyday life in advanced capitalist societies, before discussing the image in various contemporary instantiations of capitalism and what that means for traditions of radical cultural practice which attempt to critically engage with the new image condition. Finally, Chapter 9, 'Hybridity,' attempts to return the various strands of the argument to its main arc, reinforcing why we need to forget photography and what can be glimpsed of new ways of regarding the image in culture.

The Reproduction of Knowledge

Focusing upon the problem of knowledge production in the academy might be considered something of a detour from the narrative of forgetting photography, but it is necessary to consider here because the generation of knowledge of photography is a central means of maintaining the contemporary fiction of photography. Importantly for the specific horizon of this book, disciplinary distinction and academic specialisation play a significant part in how photography and its relationship to representation in and of the world is 'reified'[6], in the Marxist sense, or taken as an autonomous ontological unit, or simply taken for granted. In the current system of knowledge, the photographic image is constituted as a relative object within taxonomic regimes and optics of attention, such as everyday life, aesthetics, collection, display, media and technology. These regimes and optics are organised within the discursive boundaries of art history, anthropology, contemporary art, philosophy, cultural studies and media and communication. In broad terms, knowledge has been commodified and functionalised within systems of information. Jean-François Lyotard predicted in his seminal paper on *The Postmodern Condition: A Report of Knowledge* (1976) that knowledge which cannot be translated into data will disappear. Since the publication of Lyotard's report on the future for knowledge, universities have been corporatised, monetised, expanded and differentially globalised. Academics have been de-skilled and stratified in a neoliberal division of labour and knowledge, both in terms of knowledge production as research, and in its dissemination as learning and teaching. Essentially, knowledge has been privatised and what has been lost is the shared and public use value of knowledge.[7] In universities, postdoctoral researchers will mostly likely work on insecure teaching contracts, with limited opportunities for tenure where an average of 40% of teaching is delivered on short-term contracts. Midcareer scholars wrestle with stark choices between management, teaching or research pathways, increasingly defined by separate contracts, whilst older academics have most likely arrived at a destination and paid off their mortgages prior to retirement. One's position in academia can be measured by a property value. As Mark Fisher observed, 'New bureaucracy takes the form not of a specific, delimited function performed by particular workers but invades all areas of work, with the result that – as Kafka prophesied – workers become their own auditors, forced to assess their own performance' (2009). This discussion may seem an overly grand and distant perspective from which to begin a book on the place of photography in contemporary visual culture and, of course, on the established view of knowledge and its taxonomies, photography would rank very low on any scale of planetary urgency. But that's the point – as long as academia and education more generally continue to approach the world through ever greater hierarchical sub-divisions of instrumentalised subject knowledge, the more the paradox of an excessively knowing world, knowing it needs to change but collectively not knowing how, is replicated.

Returning to the context in which the book has been produced, whilst it is still arguably the case that universities, however differentially, remain home to independent critical thought, Lyotard's questions about the effects of datafication and audit metrics on universities still stand. The commodification of knowledge also presents the problem of how to reconnect critical knowledge or scholarship of a field to the practices it speaks of and to the world. What is specifically at stake in knowledge of contemporary visual culture is the need to bring critical understanding in theory and practice about the image back into the world and into a collaborative, transdisciplinary field and conceptual framework, something which has been previously achieved for cultural practices at a number of critical historical conjunctures. Stuart Hall and Doreen Massey (2010) usefully refer to cultural conjunctures[8] to identify a constellation of opportunities in which formal education met with a larger desire for knowledge and understanding, such as the start of the Open University. The task of achieving a unified and accessible knowledge and understanding of the contemporary image will therefore involve disciplinary knowledge translation and recalibration in order to develop a new common vocabulary about contemporary ways of seeing. This can be achieved through shared criticism and scholarly review, but it also needs to be carried out in wider public contexts and across a broad range of cultural practices. Essentially, the task is to achieve a new public educational perspective on the place of the image in communication in the age of the Internet and computation. How to do this is, of course, not simply a matter of identifying the need, but many elements are already at work at many levels of cultural communication and some kind of inventory of current initiatives might be needed. The challenges of this project can therefore be summed up as establishing more than a degree of epistemological critical self-reflexivity in setting out a conceptual frame of reference. To overcome discipline boundaries and preserves whilst not falling back into or privileging a preferred disciplinary position. To make productive and playful use of the obvious paradox of offering what is inescapably another critique whilst claiming a position of post criticality. To direct the analysis to positive and practical possibilities and to make forgetting photography a productive exercise.

2

Zombie Photography

A paradox, as it is unfolded by Augustine the rhetorician: how can we speak of forgetting except in terms of the memory of forgetting, as this is authorized and sanctioned by the return and the recognition of the 'thing' forgotten?

Paul Ricœur. 2004. *Memory, History, Forgetting*. Chicago, IL: University of Chicago Press, p. 30.

First, we can say that the analogue photograph, the world to which it belonged, the world it showed us, together with its apparatuses, although not its cultural institutions and archives, have all but disappeared. And with the disappearance of the material, technical and industrial organisation of analogue production, photography's cultural authority, vested in the accumulated knowledge of its practices, is now threatened with redundancy and finally obsolescence by a new mode of production and reproduction and its allied technologies. Second, in what is still taken as photography (we need a new term for it, but provisionally we might call it the network image), it is computational systems which now define and shape the cultural default of visuality, which is busily establishing new relations of subjectivity on the one hand and what is constituted as external on the other. But in everyday experience it is hard to see and grasp this radically new situation, in part because of the persistence of photography, which masks the changes taking place.

Photography as we persist in referring to it is dead, it is a corpse, but one which refuses to die; it remains an animated corpse and in cultural mythology it would acquire the status of a zombie, caught paradoxically between life and death. Calling up zombification to characterise photography in the twenty-first century is less fanciful than it might at first appear, as the academic literature on the zombie myth attests. The zombie hovers been life and death, as an allegory and metaphor for alienation under capitalism, expressed in terms of capital's insatiable appetite and mindless greed. The zombie is a paradoxical symbol of the fundamental irreconcilability of capitalism and humanism (Lauro 2017). Zombie scholarship traces the modern zombie of Hollywood to colonialism and to Black subjectification by White religion and to the slave system of which Haiti was an epicentre. The alienated labour of industrial capitalism was preceded by the capital accumulation made possible by slave labour. Christopher M. Moreman and Cory James Rushton (2011) distinguish between the Haitian religion and practices of what they define as Voudou and its appropriation in Westernised popular culture, signalled by the term Voodoo. In Voudou, the zombi astral is a sufferer, a captured soul, whereas the Hollywood zombie is a toiling drone of the slave plantation.

The forces which create zombies, whether those of historical cultural belief or its metaphorical application to the forces of capital, explored by cultural scholarship, relate expressly to a political analysis of who decides who lives and who dies. Power over life and death has a wider theoretical reach in the concept of necrocapitalism[1] (Mbembe 2019), which is discussed later in this volume in considering zombie photography's agency in post-capitalism. The use of the term zombie to characterise actual forces at work in contemporary societies brings us closer to its application to photography. Henry Giroux (2010) compares the fascination with the zombie in US popular media culture to a zombie politics of neoliberal capitalism and its mega-corporations, which cannibalise the economy and destroy social justice and human rights. For Giroux, zombie politics limit the discourse of freedom, substituting a zombie language that hides unjust and repressive power. Here the figure of the zombie not only devours democratic gains, but functions as a linguistic veil obscuring the disparities in human conditions. Ulrich Beck uses the phrase 'zombie categories' in laying out his analysis of capitalism's reflexive modernization (Beck and Beck-Gernsheim 2001). He argues that the categories of state-centred power, domination and politics, taken as a given in neo-realism and social science, are not capable of grasping the new situation – a situation defined by Beck, Anthony Giddens and Scott Lash as post-traditional society, ushering in a new quality of power struggles, new actors, new strategies and their associated social and political upheavals, paradoxes and ambiguous prospects (2007, p. 63). How could photography not be infected in some way in such a radically altered economic and social landscape of the twenty-first century? It is not beyond the bounds of reason to expect that photography will also have a zombie double, a manifestation of instability and alterity in the paradoxical present.

It is to the zombie as a paradoxical figure that the state of photography is attached. A conception of photography trapped between the past and the present, rendered in an altered body, but ceaselessly devouring subjectivities, roaming reality and preying on the human loss of identity and insecurity about the future. A different twist on the zombie metaphor is applied in the field of media archaeological studies by Garnet D. Hertz and Jussi Parikka, who say that zombie media is that which is not only out of use but resurrected to new uses. In zombie media, media never dies, but rather, 'decays, rots, reforms, remixes, and gets historicized, reinterpreted and collected' (2012, p. 430). Applying this logic to photography opens up the recognition that zombie photography is not simply the undead material technology of the analogue, but also that it has a remixed digital double. The zombie of photography is not the technology, which itself is relational, but a received and embodied set of ideas and practices standing over and pursing another set of objects and images. This volume goes on to identify and partially investigate three specific examples of the zombie form, that of the photographic document, of photographic art and of photographic digitisation, but in all probability there are more. Such zombies continue to roam our cultural and educational institutions as well as invade the hybrid networks. The examination of the zombie form is further elaborated, as outlined, in Chapters 5 and 6, in the case studies of the Tate galleries, the Victoria and Albert Museum and the Google Cultural Institute. Join the resistance to the zombie discourse of photography!

Forgetting

In the present, where photography continues to appear in a corporeal form, infecting the current modes and conditions of the image, the strategy of forgetting photography is a necessary antidote. Forgetting photography is a polemic, but to be understood neither as an act of wanton vandalism nor as a careless intellectual act of misplacement. Forgetting, as it is used here, is not an attack on the reliability of memory; it is not to be taken as a physiological weakness, but rather a disturbance or reorganisation of historical and collective memory. The aim of forgetting photography is not to consign it to oblivion, because the struggle against forgetting, the injunction not to forget, is an equal force calling up the reserve of recollection. Forgetting photography is proposed as a productive method of placing photography within the historical archive in order to articulate the new conditions of the image. Forgetting photography can be framed within a discourse of memory, involving remembering as well as forgiving. For Paul Ricœur the problem of forgetting is the problematic of memory and faithfulness to the past, whilst the problem of forgiving is that of guilt and reconciliation with the past. Ricœur designates the intersection of their respective itineraries as an event horizon in which a memory is appeased. In phenomenological and epistemological terms the afterlife of photography can be understood as an event horizon, and hence be forgiven as a happy forgetting and remembrance.

Ricœur's larger project in *Memory, History, Forgetting* (2006) is to chart and interrelate the phenomenology of memory, the epistemology of history and the hermeneutics of historical circumstances. He does this ultimately in order to confront the problem of the representation of the past and the enigma of the image, both in memory and imagination, which stands in the place of what is absent. This leads him to speak of the empire of forgetting, a divided territory torn between the threat of effacement and the reassurance of the reserve of remembering. What is involved in forgetting photography is an examination of its mode of reassurance, its overpowering familiarity; it is a question of how its historical epistemology and its representational image is exorcised. Forgetting is not the enemy of memory, but its process of (re)negotiation. As Ricœur asks, 'Could a memory lacking forgetting be the ultimate phantasm, the ultimate figure of this total reflection that we have been combating in all of the ranges of the hermeneutics of the human condition?' (2006, p. 413). Surely total reflection is another version of the zombie. The strategy of forgetting photography will not be easy and entails seeing precisely how what is now a fiction, or, better, an altered state of photography, is maintained and reproduced by its own discourse and knowledge practices.

The approach to forgetting photography taken here is an exact reversal of the position Roland Barthes comes to in *Camera Lucida* (1984) when asking himself what photography is in itself. Barthes's late ontological turn to the essence of photography (this is not the Roland Barthes of *Mythologies*) leads him to consider the famous absent photograph of his mother, from which he draws the conclusion that the essence of a photograph is that which has been, which he concludes to mean every photograph is a death. *Camera Lucida* expresses Barthes's love for his mother and his grief at her loss through the prism of photography. He also adopts a strategy of forgetting in the following terms.

> From the beginning, I had determined on a principle to myself: never to reduce myself as subject, confronting certain photographs, to the disincarnated, disaffected socius, which science is concerned with. This principle obliged me to "forget" two institutions: the family and the mother. (1984, p. 74)

The 'disaffected socius' he speaks of draws a line between the affective relationship of love and companionship and its objectification as society in historical epistemology. Following Ricœur, the forgetting principle proposed here seeks to forget photography 'itself', in order to see the 'disaffected socius', in order to see something of the constitution of the idea of photography. It is precisely the disaffected socius which needs to be remembered in photography, the institutions of capital and the state, within which photography and life are manifest, ordered and disordered, and in particular the institutions of culture, work, family, education and communication.

Remembering

It is the disaffected socius, deliberately forgotten by Barthes in order to approach the essence of photography as affect, to which Jonathan Beller (2018) returns in his critique of *Camera Lucida*. Whilst acknowledging the context of grief in which Barthes wrote his last work and the power and insight of his earlier semiological analyses, Beller takes issue with what Barthes understands as the 'that' of the indexical 'that once was'. Beller lays out a brilliant and compelling analysis of the racialised formation of photography and its imbrication in slavery and concludes that, 'the history of visuality needs to be entirely rewritten in terms that understand the intersecting roles of racialization, feminization and commodification' (p. 113). Beller's analysis is predicated on an understanding that slavery was systemic to colonialism and capital accumulation, in which the supremacy of the White gaze was foundational and racism naturalised. In relationship to photography, as Beller says,

> If the making of whiteness and blackness is mediated by the dynamics of photography, then the reverse is also true: the making of photography is mediated by the dynamics of whiteness and blackness. Photography does not evolve in a vacuum; it is, to borrow from Stephen Heath, a dispositif, the social and technical as photography. Thus we may expect to find that 'race relations' – that is to say, forms of racism – may be not only at the heart of 'the meaning of sight' but inscribed in the technological platforms that enable sight and, therefore, in 'photography itself.' (2018, p. 100)

Understanding the socio-technical as an historical formation of the regulation and distribution of power and its modes of subjugation is embraced by the project of forgetting photography, at least the zombie apparition of photography and its history, which has fetishised technology and enlisted it as part of the modernist project. Slavery, argues Beller, appears in *Camera Lucida* as supplementary to the photograph; slavery is referenced in order to make the essence of photography appear, but is itself disappeared. In *Camera Lucida*, Barthes uses photographs of slaves, subalterns, different bodies to reveal the essence of a technology, but in doing so represses the racialised objectification of bodies upon which the indexical image is based. The social formation which produces whiteness and blackness is written into the photographic index, defined by its spectrometry and inscribed on the surface of every photograph. Photography, argues Beller, does not, as Barthes has it, index the past, the 'that has been,' but is profoundly embroiled in the past – all of it. Beller considers all media technologies as operating in the silence of social difference, in which the social formation of an apparatus is suppressed. He notes Jonathan Crary's view that nineteenth-century optical devices were not invented in cultural vacuums and that every image technology presupposes and ideal viewer.

The argument for forgetting photography resonates strongly with current interest and approaches to the decolonialisation[2] of knowledge practices, and to critical historical counter-readings developed, as Beller notes, by 'black, minoritarian, queer, of color, subaltern, Marxist, feminist and Global South scholarship' (2008, p. 99–104). Forgetting photography opens out into a productive method for new histories of photography. From Ricœur's point of view, the historical epistemology of the recorded practices of nineteenth-century photography demand a process of remembrance and ultimately forgiveness for its violent categories and colonial foundation.

Probing the ontology of photography in terms of its epistemic framing opens out a very different vista on the history of photography. The previous life of zombie photography, when photography was still very much alive, is revealed as a much more troubled and troubling account. If in Beller's unbending terms photography could be slavery, then what else was its techno-objectification and fetishised image embroiled in? The history of photography, thought of as inseparable from the world of its operations, leads inevitably to the conclusion that if photography could be slavery, then photography can also be capitalism, the particular form of capitalism that followed the capital accumulation of slavery. This is the capitalism of ceaseless, unbridled acquisitive expansion and accumulation, a capitalism which over the course of the twentieth century, under the banner of social, economic and technical progress, has nevertheless ravaged the earth, reducing biodiversity and killing and dispossessing millions of people. Such a characterisation of capitalism could easily be applied to photography, which from the 1880s expanded rapidly across the globe, bringing everything within its purview. Whilst photography as capitalism might appear useless as an all-encompassing abstraction, it has a very concrete correlative, not only in all of the ways photography was industrially and militarily organised, but also in the prescribed world it showed. As Walter Benjamin noted, in *A Short History of Photography* (1931), in an acknowledgement to Bertolt Brecht, 'that less than at any time does a simple *reproduction of reality* tell us anything about reality. A photograph of the Krupp works or GEC yields almost nothing about these institutions' (p. 24). For Benjamin, photography freed from its nineteenth-century physiognomic, political and scientific interests becomes what he terms 'creative', and so, like Brecht's point about the image of the factory being unable to reveal the reification of the human relationships to labour, photography is limited to the surface of appearances, to fashion. As Benjamin says,

> Therein is unmasked a photography which is able to relate a tin of canned food to the universe, yet cannot grasp a single one of the human connections in which that tin exists; a photography which even in its most dreamlike compositions is more concerned with eventual saleability than with understanding. (1931, p. 24)

Much the same view of the political limits of a history of photography, which treats images as externalised and closed documents, is taken by Ariella Azoulay, who argues that photography cannot be understood separately from the catastrophes of the twentieth century. Her important book *The Civil Contract of Photography* (2012) focuses upon the situation of stateless Palestinians to describe the power relations that make photographic meaning possible. She defines a photograph as a social encounter involving several protagonists both in and out of the photographic frame and hence a form of sovereignty, which governs those photographed and those who view the photograph. Azoulay argues for a political ontology of photography which rejects the governing organising principle of spectatorship, calling instead for a civil, post-sovereign way of thinking. The formation and development of photography has been inextricably linked to the policing, persecution and violation of peoples of the world across three centuries now. The physical apparatus of cameras developed like weaponry, to shoot and capture. As Azoulay says, a violent past cannot be forgotten, but its memory can be transformed into a way of imaging different forms of life. This is memory understood in the same way as Paul Ricœur's event horizon, in which forgetting and forgiving meet. Azoulay's analysis has the benefit of leading to a practical politics of the uses of photographs and photographic archives in which spectatorship is reconceived as an act of participation in the photographic event. Significantly it suggests practices which are essentially collaborative, as her work with Susan Meiselas and Wendy Eward demonstrates, as well as being suggestive of the possibility of a collective practice of photography. What would a collective[3] history of photography, as well as history of collective photography, look like from the vantage point of forgetting photography? For Azoulay, it would 'enable one not to see the photographed persons as "governed", not to conceptualize our own "being-governed" as spectators through the regime disaster that befell them, but rather enable one to perceive them according to absurd categories such as "displaced persons", "dispossessed" or "refugees" (categories which serve external appendages of the democratic regimes under which we live)' (Azoulay 2012). A collectivist view of the history of photography would also include a reflexivity towards the ontological project itself and the difficult task of recasting the object of photography in terms radically different from its prevailing individualist and objectifying discourse and canon.

The memory of photography is umbilically connected to the photography of memory and the question arises of whether we can consider one without the other and what happens if the cord is cut. David Bate's article 'The Memory of Photography' (2010) plays with the ambiguity between photography as an outmoded technology, to be forgotten/remembered, and as an ongoing apparatus for human remembering. For Bate it is precisely what he calls the 'digital double' of photography which intensifies the interest in and concern over the proliferation of and access to historical and personal photographic archives and their relationship to memory. This clearly has been the case and he is right to ask how the new computational apparatuses of access to networks of images relate to existing notions of memory and photography. The paper discusses photographic memory in terms of Sigmund Freud's idea of a screened memory, a childhood memory which hides another, and Barthes's idea of the studium, as the averaging cultural effect of a photograph, and the punctum, the unique and subjective response to a photographic element. Bate retrieves photography from redundancy through the notion that photographs form a meta-archive of visual memory, which can be critically read as memetic traces, as prosthetic memory and as screen memories. Bate articulates the view that the photograph provides a scene in which memory can be performed, a view shared by a wider body of scholarship, which uses photographic archives to investigate human memory, especially in relationship to remembrance of the Holocaust and migration. However, what gets eclipsed in Bate's interpretative analysis of memory and photography is the memory of photography itself. Photography, however prescribed its form and however mediated its images, was not an innocent bystander in the historical events it has performed. For this reason there is a need to ask how the memory of photography can be constituted, not as something which has mutated into its digital double, but as something which belongs to the past and which has been replaced by something radically different. In doing this it is possible, analytically at least, to distinguish between all those uses of photography in the service of memory work and the memory of the work that photography performed.

Remembering photography involves stepping into its ruined territory, a metaphorical and historical space in which the available map plots a central canonical topology and selective social history, but which also contains marginal areas, roughly sketched and considered peripheral. It is in the margins of the photographic canon that a radical politics of photography can be discerned. From the preferred space of the memory of photography, its canonised and consecrated historical territory is bound together with that of its marginal critical dissenters, by that which is named photography. Twentieth-century socialist photographic projects inevitably share with the bourgeois canon twentieth-century modernism, its mode of production, distribution and consumption, drawing upon its epistemological and aesthetic techniques, styles and forms. How could the critical project do otherwise in engaging with representation?[4]

Discourses

As Ya'ara Gil Glazer notes in 'A New Kind of History? The Challenges of Contemporary Histories of Photography' (2010), the emergence of photography as a field of historical study over the past four decades is riven by two opposing forces. On the one hand photography has been defined within the discourse of high art, reflecting the increasing markets for collecting photographs and their acquisition by museums, and on the other hand photography has been framed as a social and interdisciplinary academic study. Glazer refers to Douglas Nickel's (2001) assessment as a situation in which, 'the field is caught between two opposing forces: one that construed photography as high art, with the accompanying aura of prestige, originality and uniqueness; the other arguing for "photography's social determination" and interdisciplinary character'. As Azoulay characterises the overall situation, the investment in the ontology of photography comes from perspectives of art or photographic discourse which are mostly 'unaware of the fact that what they draw as the object of discussion results from the specific field of discourse of which they are a part, and which perceives the photograph strictly in its own terms' (2012). More fundamentally, constructing a history of photography as an aesthetic medium of individual practitioners reduces the apparatuses of photography to one of an expressive tool and in doing so once more conceals the socio-technical apparatus of photography and its civil contract.

What is fraught and paradoxical about the project of maintaining the discourse of photographic history and theory is that it sets itself the impossible task of purifying the photographic image from its hybrid entanglements with the world. In practice the process of purification[5] that takes place in the photographic canon is one in which the photographic object becomes interchangeable with the abstract object of the discourse itself. Such a situation is present in the most enlightened art historical perspectives, as evidenced by Geoffrey Batchen's intervention in his 'Proem' in the May/June 2002 issue of *Afterimage*, Volume 29, Issue 6. Batchen's list of demands for a new history of photography is in itself beyond reproach in calling out the colonial, patriarchical, imperial and teleological architecture of the photographic canon. 'Proem' set an extensive agenda for historical revision within photographic scholarship, demanding a history that is true to life, complex, rigorous, sceptical and acknowledges subjectivity. Moreover, it needs to be a history aware of its own power, which is not afraid to call itself a politics and which 'acknowledges that photographs have multiple manifestations and are objects as well as images'. Batchen's call asks for a revised history of photography, which needs to think differently and go beyond the borders of Europe and the United States.

In the eighteen years which have elapsed since Batchen's 'Proem', photographic scholarship might be seen to have acted upon his manifesto in a number of revisionist and ultimately conservative ways. On the one hand a number of scholars have attended to the question of what kind of history is needed, whilst others have worked on extending and enlarging and reframing photographic histories. Batchen's point about what kind of history is needed is a political as well as disciplinary problem and one which calls into question the parsing of photography into different discursive fields. As long as disciplinary interest rests on photographic historiography, then one of Batchen's other demands, that a non-linear history of photography is needed, cannot be met, because photography remains the central thread, rather than, say, visuality, media, communication, the very politics of life. Photography is once more the obstacle to understanding the image differently. On the second question, of reframing the geographic and political borders of photographic histories, the problem, as Tanya Sheehan points out in *Photography, History, Difference* (2014), is that reframing can simply become a long list of additions to the canon in ever-expanding attachments to the existing historical structures of thinking, rather than, as she calls for, rebuilding their foundations from the bottom up. So, the paradoxical nature and continuing vexations of (un)thinking photography are endemic as long as the paradigm of photography is retained for the historical continuity of visual representation and its relationship to reality. To go beyond photography, or to break with its logic, any new history will need to be written from the point of view of the future, in the future, in which the category of photography itself has to be remembered, rather than being the organising principle around which the thinking is done and historical memory is framed.

Histories

It need not concern this account for too long but it is nevertheless instructive to identify the photographic typologies contained in histories of photography in order to glimpse the reality they brought into being and to recognise what they occlude as well as exclude. It was argued at the outset that not only do we need to overcome the ingrained language and taxonomies of photography in order to clear the way to understanding the present conditions of the image, but also to overcome the limits of the reality constructed and contained by photographic representation. Published histories of photography are the additional instrumentation of the naturalisation of photographic reality. The standard reference point against which many histories of photography have been written remains Beaumont Newhall's *The History of Photography* (1984 [1949]). The book defines a foundational taxonomy for photography in terms of its aesthetic and technical evolution and the roll call of its most eminent practitioners. This is a work conceived within the art museum and from the perspective of art history. It foregrounds MOMA's key historic role in shaping photography for art museum collection, conceived as it was as the exhibition catalogue to MOMA's first comprehensive retrospective exhibition in 1937.

The power of institutions to promote cultural practices and in this case a national idea of photography is also evident much later in Michel Frizot's *A New History of Photography* (1998), which was supported by the Arts Commission of the Centre National du Livre, in the acknowledgement, shortly after the 150th anniversary of 'the birth of photography', that there was no French authored history. The large-format book, running to over 700 pages, is, like Newhall's, copiously illustrated and arranged as a chronological account moving back and forth between aesthetics and science. The book emphasises the technical and formal dimensions of photography over and above the social or anthropological, with such headings as 'light machines', 'automated drawing', 'transparent medium', 'portrait', 'speed', 'the moment', 'body of evidence', 'metamorphosis', 'style', 'beauty', 'looking at others', 'sensitive surface', 'forms of looking' and 'ritual and customs'. Like Newhall's institutional position, Frizot's project can be seen in terms of the legitimation and institutionalisation of photography. A third volume to consider is Mary Warner Marien's, *Photography, Fourth Edition: A Cultural History* (2014), which, although it expands the categories of practice to be considered in a photographic history, nevertheless follows Newhall in a chronological periodising of photographic development within which aesthetic, formal and social themes emerge. Warner Marien's account is broader than Newhall's, as one would expect of a cultural history looking back over the twentieth century. It is more global in scale and social in reach, organising photography around major events as well as social themes. It constructs a panoramic view across two centuries in which photography is a constant witness to war, criminality, social change, suburbia, technology, children, feminism, science and society, self, beauty and culture, but with no analysis of the logic and links between them. The book is also organised around photography's formal developments within an art context, covering naturalism, pictorialism, nude, portrait, movement, futurism, revolutionary art, photomontage, the Bauhaus, Dada, Surrealism, advertising, documentary, family pictures, body, fashion, art, the medium and post-photography. It is a long list of categories through which photography is naturalised as a medium and detached from the social reality in which it operates. Finally, Naomi Rosenblum's *A World History of Photography* (2019), now in its fifth edition, is no exception to the rule for histories of photography, including nearly 900 images and structured chronologically and thematically through the categories of aesthetic, documentary, commercial and technical photography and personal artistic expression.

The publishing investment in these books ensures a continuing claim over history but also their duration in history. First published in 1949, the fifth edition of Newhall's history was published in 1982. Rosenblum's history is published by Abbeville Press, founded in 1991, which describes itself as an independent publisher of 'timeless fine art and illustrated books'. Mary Warner Marien's history is published by Laurence King, which describes itself as 'a world leading publisher of books and gifts on the creative arts, acclaimed for their inventiveness, beautiful design and authoritative texts'. Michel Frizot's *A New History of Photography* was published by Konemann Books, founded in 1993 and taken over in 2003 by Tandem Verlag, a German company founded in 1994, but continues the Konemann imprint. Taschen, a German art book publisher founded in 1980, has two photographic histories in print, *20th Century Photography* (2012) and *A History of Photography: From 1839 to the Present* (2012; a selection of images from the George Eastman collection). In their pictorial purpose such books continue to define a canon in terms of a selective visual index, which acts as the significant reference for the existence of a photographic history in the same manner as other selective canons in art, design and architecture.

These periodic attempts at reforming and revising the discourse, or the boundary object of photography, are now outstripped by the acceleration of culture, knowledge and technology in commodity capitalism. Attempts to hold on to the category photography, by fashioning new links in an historical chain of an overarching idea, miss the point that photography is now a memory and that images exist in a multiplicity of relations governed not by any linear account of time, nor a singular apparatus, but rather by chrono-reflexivity.[6] A condition of commodity capitalism is that the timeline of the future has shrunk to that of the perpetual present, and therefore time is no longer experienced as progressive, but managed. The tendency to maintain the discourse of photography by revision is evident in the debates and discussions which continue to take place regarding post-photography, a defining term for the extension of photography in the present. What remains to be accounted for here is what the memory of photography invokes. It is typically put that photography is a witness to historical events, as if such events unfold of their own accord in time and space, and photography is simply in attendance, standing outside of the event, making a record determined by the intentionality and contingency of the photographer. However, photography cannot be considered an innocent bystander to history, nor the subjective interpretation of the photographer; rather, photography needs to be understood as an active agent in the making of historical scenes, its witnessing being complicit with that which is seen. Photography brings to the scene that which has already been seen, as well as that which cannot, and hence the history photography witnesses is also the history of a dominant cultural way of seeing.

Both Beller and Azoulay, in their separate ways, call for a radical reconceptualisation of the history of photography, to be thought of in terms of its complicity with the objects and events it registers. From the family snapshot across the range of photographic genres to its scientific uses, social, economic and political relations of power are enacted as well as formed by photographic practices and their archival typologies. Through the taxonomies of photographic archives as well as in private and national collections, it is possible to identify the logics of power through which the world has been thought and ordered. The archive has the capacity to reveal particular formations of historical cultural value and in forgetting photography further reveals the continuing archival logic of its encyclopaedic dream and hopelessly paradoxical project.

If photography has been embroiled in the things it represents and is therefore inseparable from the shaping of events it manufactures, what world has the history of photography shown us? How has the history of photography taken part in the co-construction of the reality of the twentieth century, what did it show, what did it conceal, wittingly or not, and what remained unrepresentable?[7]

Representation and the Archive

From the perspective of forgetting photography, its memory is to be sought in the archive it produces but so too is the memory of the twentieth century. The photographic archive is structured by an overarching narrative of technological and social progress, in which the photographic apparatus is entailed in historical development. The history of photography is built upon an evidential historical and temporal narrative, which connects the individual, society, nation, state, art, science and nature. As Allan Sekula has observed, 'photography is modernity run riot' (1978, p. 4). Modernity is the meta-narrative of the twentieth century par excellence in which a progressive, human-centric future would prevail. The objectifying realism, Cartesian rationality of the camera and expanding capacity of the reproduction of the image gave photography a selective stage upon which the twentieth century was performed, notwithstanding its eclipse by film and video, which added sound and movement to the analogue. Photography was entailed in forging new categories of human life in sports, leisure, education, industry, farming and health. Photography was enlisted in new categories of natural exploration and discovery, social reform in documented urban deprivation and improvement in housing, education and health. It delineated the characteristics of national cultures and reinforced the importance and safety of the nation state. Photography has participated in undisclosed violent and atrocious acts continuously since its widespread dissemination in the nineteenth century, but the canonised photographic image of European history in the twentieth century is one which shuffles racism and genocide off the stage, in which progress was inevitable, wars were just and unavoidable and combat was heroic for the righteous victors.

The historical work of photography can be expressed in terms of the relations between the labour and capital involved in photographic image production. This is not to collapse photography into some determinist or reductionist second-order function of capitalism and the state. The debate has long accepted Allan Sekula's (1981) insight that 'the social practice of photography is no more a *reflection* of capitalist society than a particular photograph is a *reflection* of its referential object'. However, Sekula believed that, 'photography is fundamentally related in its normative way of depicting the world to an epistemology and an aesthetics that are intrinsic to a system of commodity exchange'. Forty years after Sekula's essay 'The Traffic in Photographs' (1981), the challenge remains to understand in detail how the epistemology and aesthetics of representation relate to a system of commodity exchange. However, the difference now is that over the past forty years capitalism's mode of production and its reproduction of the relations of production have accelerated to the point where there can be no taken-for-granted historical continuity of the critical project beyond the ruined territory. Across all of the categories of events where photography was present and occasioned, its role was overwhelmingly to say: this is how things are, they could not have been otherwise, this is the natural condition of human progress. Images of the development of the increasing application of machines to human labour and in everyday life were recorded by photography as a general good. Technological progress was good, exploration was good, human affluence was good, the relief of poverty was good, the nuclear family was good, consumption was good. All of this was the surface, coded cultural images, which were not the events, whose relation to space, time and event was oblique, mediated and for many abject.

What the history of photography of the twentieth century didn't show, couldn't show – although a few radical attempts were made – were the consequences of systemic racism, sexism and genocide, together with class oppression, upon which Western democracies were built and upon which the organisation, dispersal and practices of photography embodied. The photographic documentation of indigenous peoples, rural and industrialised workers and their communities, of working-class life and of the poor, however humanist and reformist in intent, were objectifications and othering. As such they reproduced inequalities of income, health and education as natural social categories, mirrored by documentation of the good life of the rich. What is retrospectively coming to light in more general ways is photography's complicity with colonialism and imperialism and its racialised logic. In stark contrast to photography's seemingly inescapable naturalising logic of the world as you see it, there have been attempts to use photography differently. From the 1930s onwards there have been communist, socialist, feminist and Black arts movements which have used photography in attempts to show an alternative to the world showed by dominant photography.[8] This has been done both within the framing of the photographic document as well as in extended pictorial forms. The great majority of this work remains unrecognised by the institutions of collection and scholarship. But alternative photography[9] was also bound by the same bourgeois social, documentary or aesthetic forms and the acceptance of the representational logic of the photographic image.

Allan Sekula taught at the California Institute of the Arts from 1985 until his death in 2013. His photographic practice was dedicated to revealing the conditions and effects of global capitalism upon labour. His writing was dedicated to the project of uncovering photography's founding myths and to the attempt to create an alternative history of photography, considered in terms of its social relations and framed by the concepts of the body and the archive (1986). As Sekula put it, 'The lingering prestige of optical empiricism was sufficiently strong to ensure that the terrain of the photographable was still regarded as roughly congruent with that of knowledge in general' (1986, p. 56). In contrast, Sekula points to a fundamental tension between uses of photography which fulfil a bourgeois conception of the self and uses which seek to establish and delimit the terrain of the other, by what he called an instrumental realism. For Sekula, photography was haunted by two chattering ghosts: that of bourgeois science and that of bourgeois art. In the same manner as Latour, Sekula saw a foundational flaw in the modernist separation of nature and culture working its way through photography. Photography could reduce the world scientifically to a constellation of knowable and possessable facts and objects. As Sekula saw it, paradoxically as it appears now, photography in the hands of the reconstructed subject, epitomised by the luminous person of the artist, 'had the historical mission of apologizing for and redeeming the atrocities committed by the subservient – and more than spectral – hand of science' (1981, p. 15).

By the late 1980s, Sekula could point out that the uses of photographic technology in both art and mass-consumer photography displayed a privatised world view, and saw this retreat into individualism as a response to the tendency of advanced capitalism to threaten subjectivity through the instrumentalised logic of the administration of daily life and the depoliticization of the public realm. Four decades later, from the perspective of forgetting photography, Sekula was marking a significant crisis point in photographic representation, what can be seen in retrospect as a proto-stage in photography's afterlife. Sekula was also describing the emergence of hyper-individualism and its corresponding condition in the paradoxical present. Sekula's artistic response to the emptying out of embodied social meaning led him to insist on the function of labour in the production of value and to confront the concealment of the representation of labour, in terms continuous with Brecht and cultural Marxism's emphasis upon forms of critical realism. Sekula's photographic practice embodied the paradoxical position of recognising the impossibility of unmediated vision, but still holding to the task of finding an ethics of visibility. Working within an art context, Sekula's aim of finding a visual means of representing labour in the emerging shape of global capitalism took the form of his epic and highly accomplished photographic work *Fish Story* (1995). However, John Roberts (2013) questions whether Sekula's ethics of visibility might be resigned to its own marginal visability within its form of critical realism, which is another way of recognising both the increasingly paradoxical position of photographic representation as well as the growing crisis of political representation. Roberts expresses this double crisis of representation in recognising that on the one hand Sekula's critical realism remains open to partisan allegiance of oppositional politics in seeking to give voice to the dispossessed, whilst on the other hand it inevitably lies dormant in the absence of a coherent radical force opposing the new forms of disenfranchisement. From the perspective of the afterlife of photography, Sekula's work is confronted by an absolute limit of photographic representation, even in its oppositional form. Not only is Sekula's work a lament for the invisibility of exploitative labour under the conditions of global capitalism, it is also a lament for photography.

How can the body of critical theory attached to the afterlife of photography at the end of the twentieth century be productively used now in the face of the loss of the public realm and the commodification of knowledge? Does critical photographic theory share the same fate as that of photography, itself a living undead, in which theory is none other than photography's memento mori, a poignant symbol in an academic mausoleum? If so then photography theory has cut itself off from the living and from concrete circumstances, no longer able to confront the contemporary image condition. On the other hand an archive of oppositional photographic practice could be researched and gathered materially and virtually, which would have much to add to understandings of dominant modes of seeing in the new condition of the image and the difficulties encountered in expressing alternative realities.

Foucault reminds us that all history is written backwards, as it were, from the point of view of a genealogy of the present, rather than an archaeology of the past. His reminder begs the question of what a current vantage point might be from which to write back to the body of photographic history and theory. It might also present an even more complex problem that there is no clear position in the present, as yet, from which to view photography's history and theory, now separated from its object, a reanimated corpse, capable of doing nothing other than pursuing the discourse of its origin. Of course photography theory continues to circulate and be reproduced in the many educational and artistic spaces in which it registers, this present volume being a further and ironic example. What view of photographic history and theory do any of these continuous, invested and embedded spaces afford and for whom? Might it also be possible and more liberating to write photographic history and theory from the perspective of its passing, from the perspective of the future? In entertaining this possibility it would be necessary to discount any straightforward position of continuity; the assembled canon of texts, photographers and photographs would have to be reassembled; the linear provenance of the photographic canon and linear time would need decoupling and new starting points for multiple histories of photography would have to be found. By seeing photography as historically eclipsed possibilities arise for forging new bridging points between what has been suppressed and made invisible by photography in the past and the new conditions of the image in the present. The passing of photography gives rise to melancholia, as is already evident in contemporary photographic exhibition, a hauntological manifestation of photography's temporal disjuncture. A decade after Mark Fisher (1993) took up Jacques Derrida's innovation of a philosophic hauntology in British cultural studies, the ghost of photography now haunts itself. No, a future view of photography's past and its still-ghostly presence, although I prefer to think of it as a much more corporeal zombie, will require ideas which decisively break, as far as any can, with the various forms and canon of photography itself.

A position in which photography is remembered would be one that articulates a finality in the time of photographic theory and its cultural recirculation, propelled by photography's many recent deaths. In taking the latter of these positions as a more likely fit with current circumstances and conditions, time cannot but intrude – the time of the past, present and future, no longer thought as time's linear arrow, but experienced as time piling up, overlapping and in multiple registers. This is the time of memory, writing and the contingency of death. The temporal passage of existence in which histories are remade and grasped intrudes upon the consciousness of writing, which in this author's case has taken place over half-a-century and in a period in which photographic theory has been developed. This creates a special effect in writing, a stereoscopic reflex, produced by rereading theory first encountered under very different circumstances from those of the present.

The trope of death in the argument for forgetting photography needs no apology in the face of the reality of mortality. The language of death permeates the discourse and documents of photography from beginning to end. Death is never very far from the photographic image, indeed is inscribed into its apparatus, but the symbolic deaths of photography are little compared to the subjugation, exploitation, genocides and killings naturalised in colonial and capitalist systems in which photography was performed.

The problem for what was photography was not wholly of its own making, nor simply a product of changing technological apparatuses. The problem for photography, as it is for many other systemic forms of cultural coding and registration, was inescapably part of a much larger and more complex change in systems of thought and systems of representation driven by the ceaseless energy of capital. It is possible to see from beyond photography that it might never have existed at all. We are living through a rerouting of fundamental systems of human organisation established over the course of the European Enlightenment. In this process the reproduction of knowledge, technological forms of communication and social democratic politics face new challenges from a disaggregated public, made up of the hyper-individual consumer. Unless we locate what interests us about photography in a wider context of the reproduction of capitalist culture, we will not understand the scale of the problem of the continued afterlife of photography. Gilles Deleuze and Félix Guattari, in *Anti-Oedipus* (1983), claimed that the zombie is the only modern mythic figure: 'The only modern myth is the myth of zombies – mortified schizos, good for work, brought back to reason' (p. 335). Much the same can be said of photography.

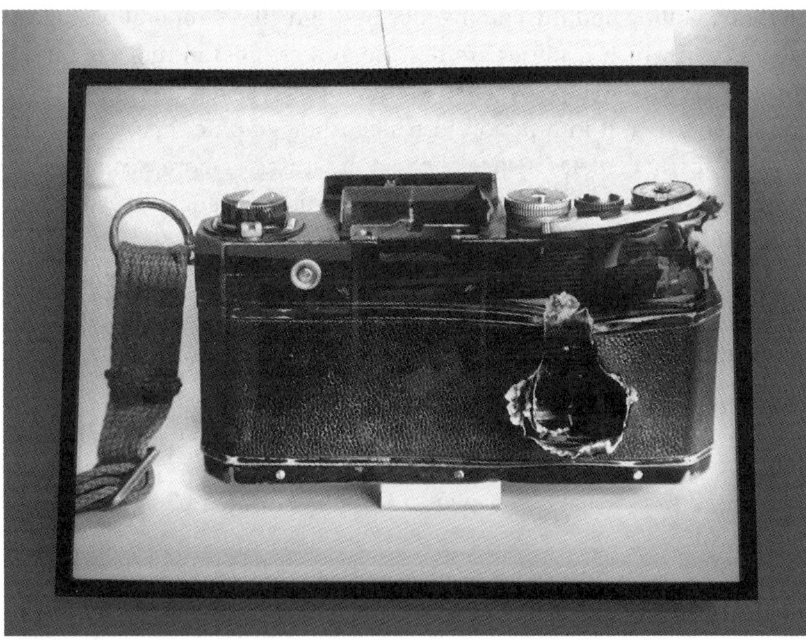

Fig. 2.1

War Remnants Museum, Ho Chi Minh City. May 2019. Author.

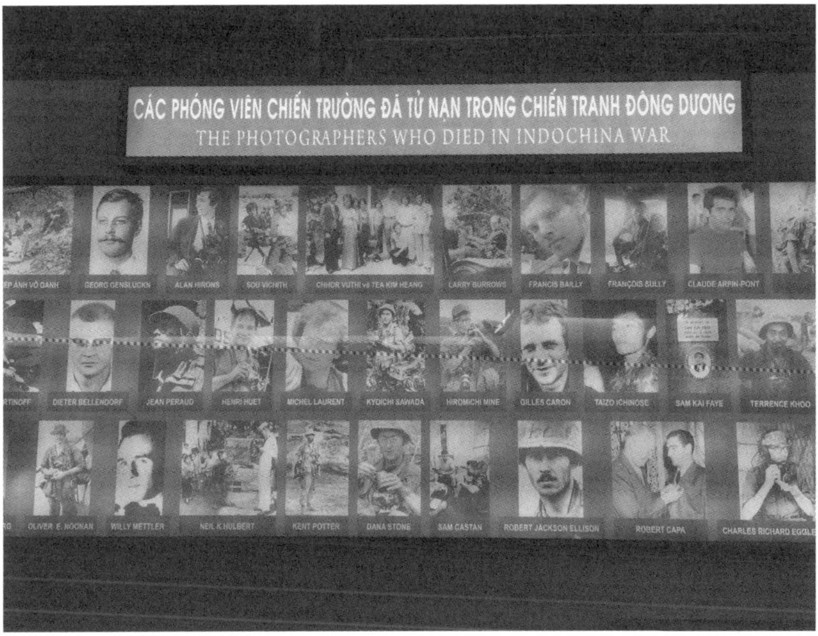

Fig. 2.2

War Remnants Museum, Ho Chi Minh City. May 2019. Author.

3

Post-Photography

We might, of course, choose to regard the digitally encoded, computer-processable image as simply a new nonchemical form of photograph or as single-frame video, just as the automobile was initially seen as a horseless carriage and radio as wireless telegraphy. Indeed, the terms 'electronic photography,' 'still video' and 'digital camera' have rapidly gained currency. But such metaphors obscure the importance of this new information format and its far-reaching consequences for our visual culture.

William J. Mitchell. 1994. *The Reconfigured Eye: Visual Truth in the Post-Photographic Era.* Cambridge, MA: MIT Press, p. 3.

This chapter considers the part played by academic discourse in which photography has been identified as a problematic object. It follows the path of attempts to situate photography within a new media landscape as well as attempts to move beyond it and is primarily interested in how and why photography remains so seductive. How is photographic theory and scholarship to be encountered as a manifestation of the afterlife of photography? The argument made here is that post-photographic discourse, rather than exploring what came after photography, ultimately maintains the historical continuity of photography. Post-photography retains the photographic image as the default of representation. It moves seamlessly from cameras to smart phones, from chemicals to bytes, in accepting photography as an embodied human practice of identity and representation. Post-photographic thinking endeavours to suture the analogue, digital and networked into a continuous image tradition in which the photographic process and language remain entailed in digital technological systems. In doing this the stance and interests of post-photography perpetuate a discourse which is radically out of step with the new conditions of the image as well as what has happened to photography. From the perspective of photography's afterlife, framing the discussion of the computational networked image in terms of post-photography is by definition something after photography, yet retains the discourse of photography. Hence the term post-photography is use here as the reprise of photography, rather than a moving beyond it, a notion designed to be annoying.

The plain argument being made is relatively simple: i) photography and technical reproduction have parted company, which presents a new opportunity to see what world photography has and has not shown; ii) scholarship, by dint of academic contractual labour and institutional competition, colludes unwittingly in maintaining the idea of photography as a current medium. As outlined in Chapter 1, under current conditions, scholars, researchers and academics experience intense pressure to specialise within smaller and smaller sub-fields. The result of myopic specialism in cultural and media studies is that the synoptic view gets harder to keep in mind, even when joining up thinking about technologies and cultural life is an important and ethical means of generating common collective concern about the future direction and regulation of media and telecommunications. The parsing of knowledge is a strong case of the wood not being seen for the trees and ultimately leads to a depoliticisation of knowledge. The demands of informational knowledge production give rise to competitive academic labour, which in turn delivers informational knowledge. This was identified at the outset of this present work and at this particular junction, where it becomes necessary to follow the details of the post-photographic thread, an engagement with discipline-specific technical terms is unavoidable. The production of knowledge of things in the world by the academy is just as much a thing in the world as the things it seeks to explain. What is structurally ruled out in scholarship is a reflexivity towards its own practice and by this means scholarship maintains a separation from the world of action. But there can be no clear separation between knowledge and its producers. On this basis it is necessary to look into how the historical narrative of the idea of photography is maintained and to do this the arguments have to be followed. The authors, works and ideas cited in the following can be thought of as the warp and the weft by which threads are woven into the fabric of a photographic world.

In the introduction to *Photography: A Critical Introduction* (2015), Liz Wells makes reference to Steve Edwards's book *Photography: A Very*

Short Introduction (2006), in which, in an opening chapter tantalisingly entitled 'Forgetting Photography,' Edwards asks the reader to imagine a world without photography. From the perspective of this book this an exciting if not crucial thought experiment, raising many questions about how visual media might have developed without photography. Can we imagine a world in which photography did not happen in the 1830s, but that television from the 1950s did? Or that it was the moving image which predominated in mechanical reproduction from the 1870s, or that computing from the 1960s could have developed popular visualisation techniques, based on sonic or heat measures, without an intermediary called photography? Could we reimagine the twentieth century in images of universal equality, fraternity and liberty? What might the intervening space of making and recording images have looked like? Worthy of note, co-editors Lisa Gitelman and Geoffrey B. Pingree, in their collection intriguing titled *New Media, 1740–1915* (2003), rehearse the importance of considering the formation of technical media and the cultural conditions in which they emerge, including many which fell by the wayside. Raymond Williams's *Television, Technology and Cultural Form* (2003 [1974]) shows in even greater detail how the eventual form of a technical medium is not a technological inevitability, but a question of political culture. In the historic case of photography, Edwards equates a world in which photography never happened with an historical loss and a dystopian future of an imageless society, although here photography is by no means the limit of the image, as the visual archive of the material world before photography, as well as the visual media that followed photography, amply demonstrate. Indeed, it should also be remembered here that the image only becomes fully synonymous with the appearance of things with the advent of photography. Objects, thoughts, memories as well as events are themselves images. It turns out, disappointingly, that forgetting photography is for Edwards and Wells a ruse, mounted only in order to achieve the exact opposite, to say that photography cannot be forgotten and the world would be a poorer place without it. As Edwards says, 'Much of our familiarity with our world comes through photographic visualisation as a surrogate for first-hand experience of places, objects, creatures and events' (2006, p. 5) and that, 'Images associated with the optical unconscious have played a fundamental role in demystifying our world; without them enchantment would have a greater grip on our understanding' (p. 6). Both of these statements suggest, as Edwards recognises, that photography came with a cost, which can be seen simply by reversing his juxtapositions, which is to say that photography disenchanted the world and relegated experience to a second order. This is reason enough to seriously consider forgetting photography, but both authors are committed to the project of preserving photography and projecting it into the future. Ironically, this is a future in which photography is already only a memory. However impossible it might seem, the task of imagining the world without photography over its historical course is what forgetting photography demands. This may read like outright prejudice against a familiar everyday object and a highly useful instrument of record, but as each stage in the argument insists the effort to forget photography might be the only way we can reimagine the multiple realities of the past two centuries of the world. The photographic record and its scopic logic of appearances still weighs heavily upon the present imagination of the material and social world. There is no photography without its objectifying frame and positivist lens and no photograph which doesn't subject and organise the viewer's gaze within a strict order of appearances, with the possible exception of photomontage, which is not at all the same thing as the interpretative experience of the human sensorium. Experience is inseparable from the material world humans are part of and which they recognise and register in multiple, affective, relational, fragmentary and contextual ways. Experience of the external world has to be thought of as shaping what that world is and can be. The photographic image, as Baudrillard might have put it, is the tyranny of the sign.[1]

Thinking about photography is shaped by recursive disciplinary practices that have become fragmented as well as muddled. Abigail Solomon-Godeau usefully points to the irony that it took until the 1980s for photography as a medium to be integrated into university departments, curricula, museum collections, art criticism and its journals, galleries and an expanding marketplace, at which point, 'in a temporal blink of an eye', photography was everywhere and no longer dependent on these consecrated spaces (2017, p. 3). From that point onward attempts to maintain photography as a discrete medium have been confined within historical or theoretical scholarship. The wider disciplinary effort in visual and media studies has been directed at the contemporary situation, reconceptualising the post-photographic in the face of the hybridity of the image and its technical apparatuses. For Solomon-Godeau, the transformed terms of photography do not herald the post-photographic, so much as the intensification and proliferation of the photography implicit in its nineteenth-century industrialisation. In this oblique reference to Walter Benjamin's emphasis upon mechanical reproduction, Solomon-Godeau is reminding us that the photographic image remains entailed in the capitalist mode of reproduction. At the same time she is reconstituting photography by locating the agency of a particular class of images as photography within image reproduction generally. Of course, the continued agency and affects of contemporary images, operating across many contexts – to register, please, disturb, unsettle and confront – has to be recognised and explored, but by the same token 'photography,' after photography, remains a settled category, comfortably ubiquitous and commonly understood. There is something somnambulant in the term itself as it is perennially re-stitched into the historical timeline.

Post-Photography Comes into View

From the 1970s, structuralist/post-structuralist accounts of photography articulated the difficulty of disentangling the idea of photography from its institutions, modes of production, reception, uses and cultural contexts (John Tagg, John Berger and Susan Sontag). The relationship between the photographic image and the reality it depicted was also recognised as polysemic, relying upon codes of language, identities and social exchange to function as representation (Roland Barthes and Victor Burgin). From the 1980s, postmodern discourse started to turn the equation of an external reality and its (photographic) representation on its head, projecting society itself as a spectacular image on the one hand and imploding the relationship between the image and the subject on the other. Through reproductive technologies and commodity exchange, the image was now seen to function in a society of the spectacle and in a reality transmuted by simulation (Jacques Derrida and Jean Baudrillard). As a response to changes in image practices and the postmodern cultural discourse, the art world and its academic counterpart in art history moved to define photography as an expanded medium. From the 1990s, with the increasing force of digital image technologies, the academy moved, in what was an interdisciplinary field of photographic studies by this time, to describe a post-photographic condition. We might sum up this selective timeline from pre- to post-photography as a number of intellectual hiatuses in and reconstitutions of the idea of photography, each of which coincide and enmesh with marked and increasingly rapid cultural change. Such moments of attenuation are visible and each expresses the central modernist paradox. As Bruno Latour has outlined, this is a paradox in which the proliferation of (image) hybrids produced by the total of human activity on and with the planet threatens to overwhelm the work of purification required in maintaining the Enlightenment constitution of the separation of nature and culture, upon which the modernist idea of photography rests.

The journey to the current reprise of photography as post-photography began, as suggested, with the theorisation of photography across the late 1970s and early 1980s and was marked by a post-structuralist claim that photography and the image of reality it showed was socially constructed, a claim which owed as much to Walter Benjamin's understanding of Karl Marx as it did to the specific theories of power and desire of Michel Foucault and Jacques Lacan. The resulting theorisations of photography, in one way or another, relativised the relationship between the photographic image and the reality it evidenced. In effect post-structuralist analysis bound photography to its institutional constitution and located the meaning of the photographic image in subjectification. Such arguments weakened but did not break the indexical hold of photography over reality, which remained necessary for the objectivist and humanist claims not only of dominant representational media holding a capitalist reality in place, but also of campaigning and worker photography attempting to reveal the truth of the exploitative nature of the ideological 'naturalised' order. Any untheorised notion of photography as an innocent, transparent medium was finally undone by the full force of postmodernism across the late 1980s and early 1990s, which assaulted all objectivist claims and dissolved society into a state of spectacle and simulation in which photographic images circulated in a system of signs, which no longer conveyed any reality other than themselves (Baudrillard). Not surprisingly this was also a moment that propelled photography into the expanded category of art and into the art market.

After Photography

The prefix 'post' of post-photography was intended to signal a decisive break with the prevailing logic of photography. The aim was to create a new space for thinking about what came after photography, in an acknowledgement that photography as previously known and described had come to an end. However, the term post-photography was also a contested, if not ambiguous, term which contained both new directions for understanding the digital condition of the image as well as continuing interest in the digital effects upon what still remained of photography. These two interests were for a time held together in the effort to understand the change from analogue to digital imaging, but eventually parted company with the exponential expansion of digitisation and online image circulation. Interest in investigating the digital condition of the image after photography eventually led to the current state of thinking about computational vision, the automation of the image, the application of AI and the datafication of value. The former interest in the effects of the digital upon photography led to a reprise of photographic history and culture under the sign of post-photography. How this happened is the subject of what follows.

Academic discussion of the post-photographic has pivoted primarily upon arguments about whether the technical development of computer imaging should be understood as a complete break, or complex continuity, with the tradition, practices and materiality of photographic imaging. Martin Lister, amongst others, was at the forefront of assessing the impact of digital technology upon photography in the 1990s, editing the influential *The Photographic Image in Digital Culture* in 1996. Lister subsequently wrote a chapter for the fourth edition of *Photography: A Critical Introduction* (2009) in which he offered a complex summary of post-photography perspectives. The book, which is widely included in undergraduate photography reading lists, is now in a sixth edition and no longer sees the need to devote a chapter specifically to the digital in what it takes as a 'post-digital' scenario in which photography is reintegrated.

The 'era' of post-photography, as Lister (2009 [2001, 2005]) charts, was initially put forward by William Mitchell in his influential *The Reconfigured Eye: Visual Truth in the Post-Photographic Era* (1992) and influenced by Jonathan Crary's *Techniques of the Observer* (1990) as defining a fundamental and irrevocable set of changes in the relationship between human vision, images and external reality. Mitchell rehearsed the implications of the shift from analogue to digital in terms of questioning the future veracity of the photographic image. However, Lister also follows the argument of Kevin Robins (1996) that whilst 'post-photography' replaces chemical photography's basis in the representation of appearances, it does so not with a loss of veracity, but with more sophisticated ways of knowing through theories, algorithms and data used to simulate images about objects and processes and concepts. This leads Lister to conclude that what is referred to here as post-photography turns out to be a continuation, on a higher more sophisticated plane, of one of the historical aspects of photography's objectivist mission, to collect facts, measure and record' (2009 [2001, 2005]).

Such radical change in image production taking place during the 1990s had serious consequences for evidence, record and value. In the early 2000s, the subsequent technical and social developments in computing, which eventuated in the networked image, were still only barely glimpsed. Some three decades on from the inception of a condition of post-photography, we are in a better position to see how the arguments and events played out. Writing in 2001, Lister rehearses a number of understandings of how photography was being changed by digital technology and postmodern ideas, as he had done elsewhere (2013 [1996]), arguing that whilst complex changes had to be granted, they did not amount to an absolute break with photography and arguments to that effect were oversimplified, or, worse, deterministic. As he said, 'The actual effects of the new technologies upon the practices of photography are both more subtle and complex as was photography's impact upon the autographic arts, and they do not include its "death"' (2001, p. 305). Lister's assessment of photography at that time was made within a tradition of British cultural studies, much influenced by Raymond Williams, which accounted for culture, here defined locally around photography, as the creative outcome of historical processes of national economic, political, social and cultural struggle for progressive change and truth. Such struggles were accounted for by Williams and broadly adopted by British cultural studies in terms of culture defined as relations between dominant, residual and emergent forms of meaning at any one given time. Williams's emphasis was always upon non-determinist historical outcomes founded upon his deep belief in a common way of life, the power of organised labour and working with the labour movement for progressive social democratic change. Not for nothing did Williams title one of his major works *The Long Revolution* (1961). In terms of understanding media technology, the cultural materialism of Williams was a reposte to technological determinist arguments, associated at the time with the work of Marshall McLuhan, which posited the historical inevitably of media forms. In the last revision Lister made to his chapter (2009), he makes the point that the four revisions he had made in each previous edition followed developments in research and applications as they took place. His updates were undertaken with scholarly detachment, given his task of assessing the cultural and technological state of photography in the context of academic debates. His conclusion at the time of writing was a measured twofold view that on the one hand digital technology is thoroughly assimilated to photography of all kinds, whilst on the other the networked image as the product of distinct kinds of computational imaging processes continues to carry with it older kinds of photography, in which computation presents an image which continues to look like a photograph.

In retrospect Lister's first survey of post-photographic debate, written at the end of the 1990s, stood at a disciplinary crossroads between cultural studies, art history and the nascent and emergent terms of new media. In the 2009 edition, Lister acknowledges a greater complexity if not personal hesitancy about the status of the photographic in mobile and computational imaging, along with the recognition that intellectual arguments about the certainty and centred place of the human, from which the world since the Enlightenment had been seen, were further complicating assertions about the value of the ways of seeing enshrined in modernist photography. The last decade has confirmed with even greater measure the limits of holding on to the photographic discourse, or the logic of continuing to argue for the photographic within computational and networked visuality.

Lister went on to co-edit *New Media: A Critical Introduction* (2009 [2003]), which articulated a cultural materialist view of media and technology. There was an inherent optimism in the project of outlining how the application of digital technologies to existing analogue cultural practices was leading to a distinct new media form. The boundaries between film, television, computer games and photography – notwithstanding the

Internet, understood at the time as cyberculture – were morphing, if not dissolving into a new medium with direct and immediate outcomes upon cultural production and consumption. New media was one of the directions the debate on the impact of digital technology on analogue media took and in important ways it avoided the question of the fate of photography, choosing instead to foreground the potential of multimedia. One of the arguments which followed from the framing of new media as a form of multimedia art was that it was television and specifically the medium of video which led critical and practical interest directed at the impact of digital technology and online culture, rather than film or photography, which remained preoccupied with the impact of the digital upon the historical cultural form and aesthetic of the image.

New Media, Art and Technology

As the technical capacity and social reach of the Internet continued to expand through the 1990s, with the advent of Web 2.0 and the development of embedded media, image sharing platforms and crucially the smart phone from 2007, the significance of the digital image expanded further, through a complex of commercial, industrial, technical and cultural investments in computational telecommunications media. The general cultural experience of the Internet in the first decade of the twenty-first century was of a burgeoning online culture. It was as if a 'readymade' new media had appeared, without warning or conventions of use. The ensuing explosion of prosumer, user-driven engagement in selfie culture, Facebook and YouTube posting, image messaging platforms and the avalanche of uploading and downloading led the academic photography community to focus upon the circulation of the image, and upon the embedded, saturated, ubiquity and increasingly mobile character of the image.

Attention to these new conditions of image production, circulation and reception spread across established studies of media, culture and communications and inevitably found expression in rethinking the cultural and technical future of television, film, print media and of course photography. However, in this radically changing media landscape, attention was also paid to the very newness of the new media and two significant sub-fields of enquiry developed early on. The first focused upon the social psychology of the Internet in cyber studies and web studies (Turkle 1997; Bell and Kennedy 2000; Gauntlett 2000), whilst the second focused upon the technical and technological character of digital media. Here scholarly attention focused upon ideas of convergence, interactivity and the immersive quality of digital image software. In these works particular attention was paid to the impact of new media upon existing analogue media in which the important concept of remediation emerged (Bolter and Grusin 2000). A fuller attempt to define new media came with Lev Manovich's book *The Language of New Media* (2001), in which he outlined the computational basis of new media in data and code. The formulations and enquiries into new media were a moment when photography finally ceased in any meaningful sense to be photography, precisely because from there on the image was produced by a nonrepresentational system of calculus and circulated in a computational network on an exponential data scale. However, as much as the technical basis of the photograph had been transformed into the computational image, the cultural basis for that image remained photography. As Manovich argues in *The Paradoxes of Digital Photography* (2006 [1994]), digital photography contains both historical continuity and discontinuity in which the mathematical code of the digital image, 'tears apart the net of semiotic codes, modes of display and patterns of spectatorship' at the same time as it 'weaves this net even stronger.' Manovich concludes by asserting that, 'the digital image annihilates photography while solidifying, glorifying and immortalizing the photographic' (2006 [1994], p. 241). This paradox remains central to an understanding of the afterlife of photography and its zombie condition. On the one hand the photographic image is simulated by a nonrepresentational code, and on the other it continues to be read through the cultural code of representation. The situation in which the digital relies upon the culture of photography has remained remarkably consistent, as recognised by Ingrid Hoelzl and Remi Marie in their book *Softimage: Towards a New Theory of the Digital Image* (2015), in which they define the networked image in terms of a relational data position, based upon photographic representation. Photography operates as the uncoupled cultural layer in the pursuit of meaning and value in the computational image, whilst continuing as the default of the representational image in post-photographic studies. The paradox pointed to by Manovich is overcome in academia by a further disciplinary parsing of knowledge of the image under the conditions of post-digital culture. Here, the neoliberal knowledge economy of university research plays its part in the continuation of an unreflexive investment in photography. The dichotomy between the proliferation of hybrids out in the world and the work of purification in the academy is temporarily resolved by annexing the image to separate functions and knowledge producing studies, rather than understood as the total world of hybrids in which the task is that of translation.[2]

The expansion of academic interest in the centrality of the image in mobile and networked communication led to a formal fracturing of attention to the image into specific and distinct frames of reference and discourse, as the demand for academic research output increased. The photographic afterlife of the image could now be constituted as so many separate, specialist research objects, to be probed and dissected in a period of academic hyperproduction in which knowledge itself became subject to the functions of datafication. Unfortunately, the same forces leading to the resurrection of photography, through the parsing of knowledge, also sealed the fate of attempts at a unified field of new media studies. By the second decade of the new century, new media as a defining term, like its predecessor, visual culture, had lost its imaginative and political reach to grasp the totality of the new conditions and was repatriated to disciplinary boundaries. Ironically the reconstitution of photography as post-photography in a conservative post-digital culture in academia and the art field was a moment of recuperation and a diminution of the larger goal of developing an overarching transdisciplinary investigation into the central conditions of the image.

As far back as 1977, Susan Sontag had noted, in parallel with John Berger, that, 'A capitalist society requires a culture based upon images. It needs to furnish vast amounts of entertainment in order to stimulate buying and anesthetize the injuries of class, race and sex. And it needs to gather unlimited amounts of information, the better to exploit natural resources, increase productivity, keep order, make war, give jobs to bureaucrats' (2014 [1977], p. 178). In a prescient insight into the complicity of photography and capitalism, Sontag argued that the need to photograph everything is part of the logic of consumption, which can never be satisfied, and that, 'attempts by photographers to bolster up a depleted sense of reality contribute to its depletion' (2014 [1977], p. 179). The same can be said of post-photographic theory. Some three decades later, Sontag's analysis has been built upon, in the enquiries by Jonathan Beller to rethink the politics of the image in the substrates of computational capitalism and by Ariella Azoulay in the politics of the civil contract of photography. Both Azoulay and Beller are interested in developing a theoretical view of the totality or systematicity of visual apparatuses in the reproduction of inequality. However, the question remains as to why the current radically transfigured status of the image and the corresponding reconfigurations of vision and visual culture are treated as relatively peripheral by the communities of photography.

The Ontological Road

Renewed interest in the ontology of photography has been shaped by widespread academic interest in phenomenology, expressed as the 'affective turn' across all cultural fields from the 1990s. The affective turn represented a turning away from critical epistemologies based upon rhetorical and semiotic systems of representation, towards a paradigm of subjectivity, embodiment and information in which the central architecture was the concept of affect. For humanities, the affective turn opened up what appeared as new political and ethical perspectives in which power was a property and affect of the body, of subjectivity and emotion. Affect allowed for a new account of the normative as a mode of subjectification, as well as creating a condition of possibility for subjectivity. For progressive and critical scholarship the affective turn was a way out of the impasse of the linguistic paradigm and that of the *Realpolitik* of Left parties after the collapse of the Soviet Union and the general failure to extend social democracy in the face of neoliberalism. For photographic studies, the affective turn produced a micro temporal focus upon the material, object-orientated agency of the photograph and camera. More widely, critical attention to the affect of objects and technologies, as Claire Colebrook has observed, was a response to the capitalist boom with its overconsumption and marketable affects, which, importantly, was also accompanied by affect fatigue. Colebrook has argued that there was an inverse relation between the wider extensions of affective stimulus and a diminishing intensity of affect (2014, p. 73). In cultural criticism the apparent paradox between the simultaneous abundance and depletion of affect became central to critiques of the overstimulating affect of rampant consumption and its consequence in the loss of cognitive or analytic apparatuses. The loss of sustained attention was also expressed as the loss of the real, as Jameson and Baudrillard, amongst others, put it, but against such past postmodern pessimism Colebrook points out that the affective turn has been both lamented and celebrated as either a retreat from judgement or as a liberation from overly calculated modes of reason. Both are reactions to changes in cultural production of which the central feature has been an intensification of technological extensions of the human brain beyond its organic boundaries.

What has been at issue for photography is how digital technologies have changed the relationship between the physiology of human sense perception and its translation into cultural ways of seeing. In this equation the 'digital photograph' has been taken as a central motif and interface function of an image-saturated world and hence the turn to the study of such saturated affect in photography's newfound ubiquity. But, in terms of where a politics of scholarly attention to affect led, Colebrook's distinction between what she calls hyper-hypo affective disorder and the abstract concept of affect is an important, if not crucial, link. Attention to the affective world of the circulation and reception of the digital image at scale and speed belongs to the hyper-affective world of emotion and subjects. Attention to the material or non-human apparatuses of photography has been framed by the philosophical concept of affect. Colebrook makes two points here; first, all of the theoretical turns are both expressive of and reactions against the glut of affect, which have led theory to insist upon the intelligence and profundity of affect; and, second, the affective turn is not a solely academic or theoretical correction to the entrenchment of the linguistic paradigm, because there is a passion for and intensity to affective consumption that is extensive, even though, paradoxically, we also recognise our inability to sense. To move beyond being subject to the intensities of affect, in theory at least, Colebrook, following Deleuze and Guattari's epistemology, takes the *concept* of affect as capable of creating new lines of thought and having the potential for thinking of forces detached from the living and from emotion. It is this abstract sense of the concept of affect, beyond the immediate human-lived responses, that has interested and propelled renewed academic interest in the materiality and apparatuses of photography, in photography's ontology.[3]

From the overall perspective and argument of this book, the reprise of the afterlife of photography has three main ontological manifestations – the pictorial, the technological and the philosophical – corresponding respectively to the disciplines which conjure them in their own image. The pictorial interest in photographic images is maintained by art history, while the apparatuses of photography are infused by the philosophical current of new materialism and practically pursued as a branch of media studies and media archaeology. Philosophy defines photography as an abstraction and speculation of thought, as philosophy in fact. A fourth disciplinary investment in photography can be found in cultural anthropology, which considers the social practices of image making, but in the terms offered here is not part of the ontological triumvirate, because its project is primarily epistemological.[4] However, in wanting to get beyond photography, through forgetting photography, cultural anthropology and visual culture, along with science and technology studies, offer a more fruitful place to consider image culture and the relationship between the image, medium and viewer.

Of course, in the endeavour to discover the ontological essence of photography, the disciplines involved are by no means absolute silos, as concepts about media practices pass selectively between them. The transfer of ideas and concepts about photography in art history and theory, cultural and media studies and philosophy follows lines of shared interest in common objects, as well as reflecting hierarchies of explanation from the most abstract to the concrete and descriptive. In this later respect over the last half-century philosophy has come to occupy a general position of theory and because of this the discussion of the philosophy of photography has been accorded its own chapter due to the length of the explanation required.

The academic triumvirate of art history, media studies (and its more recent sub-field of media archaeology) and philosophy is responsible for attempting to resuscitate the living corpse of photography through an ontology of intensive care, to work against the grain of photography's ultimate demise and in doing so misses the opportunity to assess the historically finite era of photography, as much as it misses the new conditions of the image. What is important to ask here is what is being defined as the photography of the post-photographic continuum and in what contexts and discourses does it have currency? Is photography always the same object in theoretical, historical, social, commercial, scientific and fine art contexts and practices? For example, is the photography of the exhibition *Masculinities* at London's Barbican (2019), which sets out to show how central photography and film have been to the ways in which masculinities have been imagined in contemporary society, the same thing as the photography of Vilém Flusser's book *Towards a Philosophy of Photography* (1983), in which he argues that it is the camera programme which determines the subject? Conventional wisdom would account for the relationship between the exhibition of photographs and a philosophic treatise on the medium as a formal distinction between practice and theory, between material objects and abstract ideas. Exhibitions and books are highly distinct, although related, knowledge practices, whose common object in this example remains the idea of photography. In the exhibition a curated display on a specific theme in a public gallery, in the book a published philosophic treatise on an image making apparatus, the relationship is parasitical but not identical. Such comparisons if extended would include a very diverse list of images and image making processes defined as photographic, because historically they rely in some shape or form upon the whole edifice of mechanical/analogue image reproduction and the social formations that emerged with and reproduce them. Over the course of the last 180 years, a diverse set of material knowledge practices, based upon image reproducing apparatuses, has congealed and hardened as the cultural object of photography that still operates today.

The Return of the Return

Recognising the diversity and multiplicity of the uses and situations in which technologically based imaging processes are present should lead out of the culturally framed historically patterned photographic discourse, in order to investigate what lies outside and beyond it, to ask what clues we might gain to the new hybrid conditions of visuality and configurations of the real. However, this is not a situation recognised by the disciplinary projects identified, where the very recognition of the new planetary scale of imaging practices leads back to ask what photography is in itself. The photographic discourse is nothing if not nostalgic. Writing out of art history and gender and cultural studies departments, Donna West Brett and Natalya Lusty, in their edited book entitled *Photography and Ontology: Unsettling Images* (2019), list at the outset the many uses of photography as evidence, communication, provocation, surveillance, a narrative of the self and as an artform. What they seek across all of these uses is a common thread, asking what it is about photography that allows it to be taken up in such contrasting and contradictory ways. Their answer takes them back to Roland Barthes's insight of the photographic image's 'analogical perfection,' the photographic index, in which, paradoxically, meaning develops on the basis of an image without a code. It is the paradoxical status of the photographic image as both 'real and contingent,' no longer put as an opposing binary, but as an unsteady unity, that provides the basis for defining photography's ontological condition as one of uncertainty. In relation to photography's historical relationship to events in time and space and to memory, it is this ontological uncertainty which makes it a revelatory medium, whose ontological character is, 'both in and of the world simultaneously' (2019, p. 2). For Brett and Lusty, it is the very conditionality of photography which opens up new vistas for research in which both the critical and felt, desire and loss, can be added to photographic theory. Photography, it is suggested, has a life of its own as a medium which continues to change. The ontological bridge to mapping photography into the condition of image hybridity is to privilege the camera as the first guarantor of the presence of analogical perfection, which leads out into the second moment of revelation. Such a move allows the radical networked hybridity of the image to be safely discussed as the digital image, a photograph by any other name.

Enquiries into the ontological identity of photography overlap with studies of media technologies where the ontological and epistemological character of the computational image, or capitalism by any other name, are expressed as a technological succession. Acknowledging the living-dead condition of photography represents the opportunity to take the concept of photography apart, to see it as an historical assemblage which has overrun its historical course. The longer the academic post-photographic discourse projects photography into the future, the less likely an understanding of the planetary politics of the image will be.

Post-Photography and Art History

The discipline of art history has been formative in shaping a history and theory of photography, just as much as photography has been instrumental in the shaping of art history. As art historian Donald Preziosi has it,

> The powerful network of apparatuses constituting the modern discipline of art history presupposes the existence of photography. Indeed art history as we know it today is the child of photography. From its beginnings as an academic discipline in the last quarter of the nineteenth century, filmic technologies have played a key role in analytic study, taxonomic ordering, and the creation of historical and genealogical narratives. Lantern-slide projection entered the field very early, establishing the formats of study, analysis, and comparison of images. (1991, p. 72)

Interestingly, here Preziosi refers to photography as filmic technologies and goes on to suggest that the art history slide is but a still in an historical movie, making cinema in his view the true foil to art history and opening up a wider historical scope on the relationship between the study of art and the study of media. But art history and photography are inextricably historically bound together and their coupling reveals something of art history's recurring discursive relationship to images and to the origins of photography in the nineteenth century. For art history it is as if the essence of photography, mirrored in art history's own foundations, could be discovered in photography's emergence, rather than in its subsequent widespread diffusion over the twentieth century. Before the advent of film, photography dominated the visual field, but thereafter became fused with the technologies of reproduction. At the end of the twentieth century, the development of digital media presented the possibility to consider the 'birth' of a new medium as well as to reconsider the moment when old media were new. In art historical terms, the current direction of ontological interest in visual technologies both old and new is to be discovered in the relationship between material practices and discursive intellectual contexts. However, this has led to a situation of an endless return of the return, in which the essence of the digital image is to be discovered in the birth of photography and, vice versa, that photography was from its inception digital. But rather than being the final nail in photography's coffin, this elliptical process propelled post-photography's claim that photography was foundational. Photography was reclaimed from its digital demise by the argument that the photographic image has always been a selective manipulation of visual elements determined by light itself, and hence has always been 'digital' (Zylinska 2017). In proving the case, attention increasingly fell upon comparisons, similarities and overlaps in the material apparatuses of both analogue and digital imaging.

Many of the established and influential academics writing about the fate of photography are art historians and work in university art history departments. As Donald Preziosi commented, by the 1990s, art history could be likened to a department store, where depending upon personal taste it was possible to pick and mix, from a menu of formalism, Marxism, feminism, psychoanalysis and a further array of post-structuralist theory, any one of which tendencies could take hold on an academic department, as well as being liberally polyvocal. Art history has had and continues to have its own knowledge foundation crisis, evidenced in an *e-flux* article by Sven Lütticken (2013), commenting that the emergence of global art heralded the end of art history as a discipline dependent on Western narratives. Lütticken quoted Hans Belting's remark that 'global art often escapes the arguments of art history, as it no longer follows a master narrative and contradicts modernity's claim to be or to offer a universal model'.

But while theoretical tendencies and fashions compete for attention, and disciplines are forced to reinvent themselves, art history continues to revolve around the principle and reconstitution of art and its objects and, in the particular line of argument being traced, in the continuing object of photography. A notable and influential example of scholarly work locating photography in a modernising art historical context is Geoffrey Batchen's *Burning with Desire* (1999), which considers the identity of photography in its proto-photographic moment before it was formally announced to the world in 1839. As such it steps out on an ontological and epistemological dual carriageway, with a mixed method of historical scholarship, an analytical frame of Foucault's archaeology of knowledge and Derrida's deconstructivism and the work of difference. From the perspective of forgetting photography the importance of considering Batchen's text is that it is an example of photography becoming more fully embedded in historical scholarship and the use of photography as a modernising object of art history. Batchen sets up his project as providing a third way from what he sees as the limited binaries of and opposition between formalist and postmodernist definitions of photography. The formalists, he notes, looked for the essence of what photography was in itself, seeking its nature within the categories of art history, whilst the postmodern argument, which erupted across a range of disciplines, defined photography as contingent and mutable in culture. Batchen sees contradictions and similarities in both positions, but particularly in what he takes as the postmodern position, which assumes that whilst the mutability of photography can be delimited by the cultural, its identity cannot. This he sees as an essentialising gesture, shared by the formalists in their insistence on nature, but unacceptable from a postmodern position, which disavows essentialism in all forms.

Batchen asks the question, 'Why, amid the general postmodern critique of binary structures, does this division between sameness and difference, nature and culture, substance and appearance, continue to be essentialized?' (1999, p. 21). This would indeed be a serious question, were it directed at what constituted the postmodern view of photography at the time he was writing, in the late 1990s. The postmodern pantheon Batchen rehearses consists of John Tagg, Allan Sekula, Victor Burgin and Abigail-Solomon Godeau, all of whose theoretical perspectives and allegiances were forged on the anvil of structuralism/post-structuralism in the late 1970s and 1980s, informed by Marx and Freud in the fires of late modernism. These were all writers concerned with the critique of capitalism and its dominant forms of power, viewing history not from the hyper-relativist postmodern positions of the end of history, the era of simulation, the collapse of the sign and the signified, nor the society of the spectacle, nor even the advent of hypermodernity, but from more localised positions of identity within progressive social democratic change. Their analysis of photography's embeddedness in social and economic systems and practices is continuous with the structuralist search for a knowledge of underlying systems of meaning, rather than from a postmodern critique of

Enlightenment epistemological grand narratives. Moreover, the particular group of writers singled out by Batchen were writing across the Anglo-American axis, as writers for whom the full force of postmodern continental theory was still breaking and the cultural movements of postmodernism still arriving. Victor Burgin's edited volume, *Thinking Photography* was published in 1982, whereas Jean Baudrillard's *Simulacra and Simulation* was first published in English in 1983.

The distinction between what is identified as post-structuralist thought and that of the postmodern is complex and while terms have been used interchangeably in academia and cultural criticism to describe the characteristics of late modern philosophy and culture, they designate different kinds of cultural phenomena. The characteristic postmodernist thought, which tore down the binary oppositions of nature and culture, subject and object, reality and representation to replace them with the terminology of alterity, uncertainty, immanence, difference and desire, derived specifically from Derrida and the work of Deleuze and Guattari. In fact the uncertain balancing act between the postmodern and post-structuralist is much more in line with the interesting journey Batchen charts in accounting for the identity of photography. If this is the case, then it would go some way to providing an answer to Batchen's original question, because the writers on photography deemed to be representatives of the postmodern were in fact pursing the end of late modern certainties which involved binary taxonomic categories and to locate photography within progressive politics. Putting the formalists and the postmoderns on one side of the scale in order to weigh another post-structuralist art historical position of enquiry on the other depoliticises the argument, by setting up an unbalanced and inexplicable view of the place and work of theory. The excavation of a psycho-social identity of photography belonged to a specific cultural formation more aligned to political activism at a time when academia had little investment in photography, which in any case was taught primarily as a technical subject. Batchen's art historical project on the other hand is a fully fledged academic undertaking, coming at a time when photography as a proper object of scholarly attention was in the process of establishing itself. In the legitimation of photography as a proper object of scholarly interest the post-structuralist/postmodern had a conservative force in an academised body of theory across the

Anglo-American humanities, well beyond its original French cultural context and the intellectual projects of its protagonists. The substance of Batchen's account of the desire to photograph returns scholarly interest to the pictorial and to the photographic archive, and ultimately, it might be argued, a return to the formalist photographic canon and most certainly, as Batchen directs, to the future interests of art and philosophy, which sit safely back in the domain of art history.

Beyond the art historical discipline, the full force of the collapse of confidence in the modern and its critique remains the driver of much of the current and protean intellectual and political upheaval across all kinds of social and cultural institutions, which in practice means that the post-modern is far from being a settled historical term or category.

In closing, Batchen looks towards the future and makes a number of remarks on the impact of digital technology upon photography, in much the same way that Lister had at roughly the same time. He noted, as others had done, in a riposte to the idea that the digital had removed photography's truth claim, that photography had always been manipulated, arguing that 'changes in imaging technologies will not cause the disappearance of the photograph and the culture that sustains it'(1999, p. 213). This is a widely uncontested view as it is not and never has been some reified notion of technology alone which brings media forms into existence, nor vanishes them at some point of obsolescence. Indeed, to think in terms of imaging technologies as if they weren't also cultural formations is now a misnomer. Batchen also adds that 'reality may have been transcribed, manipulated, or enhanced, but photography doesn't cast doubt on reality's actual existence' (1999, p. 212). This is a more contentious and somewhat surprising statement given the alterity and conditions of possibility ascribed to photography, and on a number of postmodern views it might be said that photography performs the existence of any actual reality. As argued in the previous chapter, the reality photography inscribed was bounded by a modernist discourse which has been shown to be deeply flawed. Two decades ago, Batchen had the positive insight to see that 'Photography's passing must necessarily entail the inscription of another way of seeing – and of being' (1999, p. 216). The time of photography's passing was already occurring and post-photography as a project of art history represents the lament, the memorial and the holding on to the memory of photography.

It is also the case that, as understood by post-human perspectives, the last two decades have confirmed new ways of being. It is now possible to attend to the past era of photography, as something not to be confused with the new condition of the image and the world. The history of photography can now be reconstructed, as Foucault's genealogy of knowledge insists, from the point of view of the present, although in this particular case it will be from the position of the future. But, as the title of this book echoes, Baudrillard criticised Foucault's too-perfect theoretical universe for reproducing the very thing it attempted to uncover, the reality principle.

Post-Photography and the Materiality of Media

Media archaeology has influenced recent approaches in photographic research and stands in, in this account, as a second turning on the ontological road to post-photography, which can be observed in scholarly papers which foreground the technological apparatuses as a means of photography's revivification. An archaeology of photography involves an excavation of its technical constitution, in which the wood, metals, papers, plastics, celluloid, chemicals, batteries and glass are held to contain dimensions of memory and time. Modifications to cameras, enlargers and processing over time, can, so the method prescribes, be deciphered as signalling code, revealing something of photography's essential nature. Such a media archaeological approach has been pioneered by Wolfgang Ernst (2012). Trained as an historian, Ernst saw in the media theory of both Fredrich Kittler (1999) and Marshall McLuhan (2001 [1964]) that media was the forgotten term in historiography and that through media, new historical discourse could be uncovered, if not actualised through an archaeology of media. For the purposes of our discussion of post-photography a limited focus upon Ernst's commentary on photography will have to suffice. Ernst's media archaeological approach importantly adds the (non-human) agency of machines to the writing of history and historical method. Ernst takes the physical property of light recording as the key material fact of photography and as such he sees it as introducing a distinction in images between describing and showing. For Ernst there is no interpretation in a photograph, only a direct transfer of light, and for this reason photography became a radical element, central to the codes of modern visuality. In making the photographic index his starting point, Ernst follows Barthes in seeing photography as a decisive mutation and rupture with previous informational economies. Ernst sees photomechanical reproduction at its non-discursive level of direct inscription, whose ontology belongs to the world of physical reality. The materiality of the photographic process and its apparatuses come first when considering what constitutes its knowledge of history. At the same time Ernst acknowledges the parallel position of culturalist and art historical scholarship, which emphasise that the technical media of photography operated in parallel with the techno-culturally interpretative pictorial and representational codes of painting, printmaking and engraving. Nevertheless, in his own account Ernst relegates the culturally discursive to a secondary position, privileging technical media accompanied by what he terms a cold media gaze, which augments the autonomy of machines whose agency is unencumbered by emotion or semantics. The idea of a cold mechanical eye leads Ernst to claim that technical media is a prior level of coding the past. Technical media are time machines encoding the noise and physicality of the world outside of human-centred sensory perception, intentions or signifying structures. In this respect Ernst is emphasising a non-human dimension of photography, which aligns with theoretical perspectives on the post-human.

Jussi Parikka acknowledges the work of Friedrich Kittler as formative in media archaeology in the broader context of German media theory. Parikka notes of Kittler, 'By "upgrading" Michel Foucault and other critical arts and humanities theoretical approaches to be equipped to tackle technical media culture, such approaches have insisted on technical specificity in terms of how we talk about culture and communications in the age of scientifically based technical media apparatuses' (2012, p. 95).

Geoffrey Winthrop-Young, who has translated Kittler, also wrote *Kittler and the Media* (2011), in which he notes that Kittler is by no means a straightforward academic scholar and theorist and that he is notoriously hard to interpret and pin down. Having said this he also estimates that Kittler has had a significant influence upon contemporary thinking about media. Winthrop-Young talks about the 'Kittler effect' and identifies three main contributions of importance to students of media and communications. The first is that Kittler provides an effective antidote against presentism and uniformitarianism in his radical historicism, introducing the idea of fundamental historical discontinuity to media studies and modelling changes of medial and communication structures on an expanded historical scale. Second, by introducing post-structuralist theory to media studies Kittler removed a longstanding naivety in the discussion on technology and in return brought an informed media technological perspective to the orthodoxy of cultural theory. Third, Kittler has provided a revised framework for the discussion, long associated with Marshall McLuhan, on media determinism and the ongoing discussions of post-humanism.

For media archaeology, then, photography's meaning is inscribed in the layers of the technical knowledge of its apparatuses, its chemistry, papers and optical lenses and the material systems of production. Such an approach represents one important aspect of the ontological turn, ushered in as an antidote to what was seen as the exhaustion of semiology and the rejection of constructivist relativities, in which meaning is defined by the socially mutable. Media archaeology is contingent with, but not limited to, the ontological turn in photography, since it also represents a much bigger project to recast historical epistemology through technical media. As an approach to constituting a new history of photography after photography, media archaeology has much to offer. Further research in the photographic archive could begin to unpack the very visuality it helped create and in doing so show us the character and limits of its representational system. It could also contribute to a techno-cultural perspective on the history of photography to provide a more nuanced and deep-historical account of photography's relation to colonialism, war and the physical forms of human rights abuse through its material practices. It could also offer new historical perspectives on photography's cultural material relationship to the nuclear family at one scale and planetary degradation at another.

More problematic, however, is media archaeology's approach to the continuing afterlife of photography. If we are already beyond the era of photography, then a media archaeological approach to the current conditions of imaging would have to part company with the idea of photography, something Dvorak and Parikka (2021) remain ambivalent about. Media archaeology acts in the manner of the cuckoo; by laying its eggs in photography's nest, it aims to supplant a different analytical engine, based upon what technical apparatus can reveal about social and aesthetic coding. Media archaeology can link the technical media of electronic imaging to the material technology of photography and media through decoding apparatuses. This is evidenced by Ernst when he says 'the photograph is an assemblage of optical signals', whether analogically or digitally registered, but also wants to say that 'In the present age, the possibilities of digital manipulation of electronic photography seem to be returning images to a prephotographic quality of painting: that characterised by the painterly brushstroke' (2012, p. 47). He also says, 'digitalizing technologies are absorbing the photographic image. Thus photography as a visual technology in its own right could vanish like the image of a face drawn in the sand at the edge of the sea (to borrow Foucault's metaphor)' (2012, p. 41). Of course the question of photography's vanishing is of much interest here, precisely because of the question of how a material historiography marks the emergence of new media and the obsolescence of an old one. On the one hand the media archaeological approach of Ernst suggests that the technical media of the computational networked image marks a profound historical rupture in the operations of visuality established by photography, and on the other hand he leaves the door open to the continuation of photography as a discursive system through its technical absorption.

In the wider context of computational automation and artificial intelligence it is not surprising that media archaeology has looked to new materialist philosophical currents as a means of redefining traditional perspectives on the relationship between media and society. As Parikka says, 'mixing philosophy with media theory offers an insight to why we are so interested in non-human bodies and objects, processes that escape direct and conscious human perception, intensity of matter of technological and biological kinds' (2012).

Media archaeology recognises that certainly electronic media have built-in obsolescence, which creates a zombie like condition through recombinations of technical knowledge (Parikka 2012). However, the method of decoding apparatuses continues to privilege technology over the social, or to only see the social read through the technology. The ongoing academic project of legitimating media archaeology within media disciplines runs the risk that instead of centring the political complexity of material cultures, it can easily become an ontological short-circuit in order to breathe new life into the living corpse of photography. At worst such approaches repeat an ever-more-complex form of technological determinism, which is far from the objectives of the proponents of media archaeology.

But here an important intersection arises for critical interest in the relationship between the new condition of the image, the real and the everyday and more generally in understanding the relationship of technology and society. On the one hand the technology/society question returns directly to how the academy develops and expresses critical perspectives upon disciplinary forms of knowledge, while on the other hand it calls into question how critical interest is lodged in anything resembling a public sphere or common culture. The questions it begs are what does critical knowledge do in the world, who is it for, where does it go and what are its effects as well as affects? Such a recognition of the problems of the constraints of the current knowledge economy, for humanities scholars, has specific implications for critical interest in computational culture's rendition of photography.

Critical interest in the long afterlife of post- or expanded photography now centres upon the ways in which the photographic image in network culture operates on the smooth surface of the screen. Or, as Nina Lager Vestberg (Dahlgren et al. 2013, p. 113) puts it, 'the technical signs of photography are everywhere programmed into the algorithmic infrastructure that supports our current digital visual culture'. The programmed signs of photography, or its simulation, has led some to see erroneously that, ontologically speaking, it is the photographic rather than algorithmic medium which carries the key to understanding the culture of the visual which is emerging. Such a view is exemplified by Joan Fontcuberta who, writing in the 2015 catalogue *The Post-Photographic Condition*, which he edited for Le Mois de la Photo à Montréal in 2015, expressed the following: 'I believe that post-photography is not a style or a historical movement but a rerouting of visual culture. It is a concept we should express serious interest in because it defines a new relationship we've adopted with our images. Photography is an instrument that allows us to consider who we are today, what our society stands for and what it will become.' In coupling the photographic with the algorithmic in an embrace of the apparatus, post-photography discourses obscure the narrowly defined, highly controlled, cruel and unjust frame of reality photography has historically prescribed and holds back a more emancipated understanding of the relationship between humans and non-humans.

The elision of analogue and digital, photographic and algorithmic, image and medium in post-photographic discourse belies a partisanship in the desire to rescue photography and hence in many respects is uncritical of the crisis of representation and the forces shaping it.

Part II

4

Philosophy, Technology and Photography

A philosophy of photography is necessary if we are to lift photography into full consciousness. To do this is necessary because photography may then serve as a model for freedom in the post-industrial context.

Vilém Flusser. 1983. *Towards a Philosophy of Photography.* London: Reaktion Books, p. 82.

The digital marks an immense transformation – if by immense we mean having no measure. The impact of this transformation is still difficult to comprehend. Apart from the fact that we lack sufficient distance to take its measure, this profoundly disruptive moment explodes all frames of thought.

Bernard Stiegler. 2016. *The Digital, Education, and Cosmopolitanism.* Berkeley, CA: Univeristy of California Press, p. 157.

As philosophers, Both Vilém Flusser and Bernard Stiegler saw a pressing political need to engage with contemporary technology in order to reveal as well as oppose the instrumentalisation of life and the drive to informational automation. They also argued that technology offered a future liberation for humanity if thought through philosophy. The following discussion of philosophy, technology and photography is undertaken in the limited context of commenting on the place of photography's theorisation within institutional knowledge practices. It does not propose any theoretical perspective, nor does it engage directly in theory, but rather asks what work theorising photography does through philosophic discourse in the reproduction of the idea of photography. The chapter does not aspire to a comprehensive survey of philosophic perspectives on photography; rather, it selects a number of examples to illustrate how the photographic continues to be entailed in the technological present, to keep photography in view, through the practice of philosophy, which is to say through abstract speculation on the ontology of the photographic.

In contrast, forgetting photography is an attempt to remove photography from view in order to make way for the technical image. Forgetting photography grapples with the paradoxical situation in which the general mode of image production is no longer founded upon the photographic and yet photography persists as a discourse of the image in the world and the apprehension of the world. Keeping photography in view increasingly obscures what the current technological image is doing and the world it is bringing into existence, which can no longer be usefully explained in photographic terms. Such a paradoxical situation is indeed a theoretical as well as practical problem and hence a rightful subject for philosophy. But what also need to be taken into account are the conditions which give rise to the abstractions of photography, which is to say the parsing of knowledge and its disciplinary, institutional and individual contexts. Questioning what photographic discourse does or produces could be thought of as an anthropology of the image, or, as science and technology studies might put it, the attempt to trace the agency of discourse within a network of objects, practices and ideas. In this sense philosophic thinking about photography is understood here as an actant in a socio-technical assemblage, or located within a social ontology, which keeps photography in view. A full account of philosophy's place within a social ontology of photography would go far beyond what is laid out here, which is more of an extended footnote upon some of the ways in which philosophies of photography have been mobilised and entailed in the wider, most often educational effort to understand the contemporary image condition. The argument put forward is that philosophy, like the discipline of art history and the field of media archaeology, currently provides a theoretical bridge between photography and new media discourse. Rather than taking the step of recognising the historical limit of photography as a technology, medium and culture, the ontological focus upon the technology of photography leads, without interruption, from the historic discourse of the photographic image into the computational mode of image production.

The links made between photography and computation are a problem in the conceptualisation of difference and of language designation. In the knowledge discourse of social semiotics and anthropology, language operates to identify, name, describe and analyse all things external to itself, as well as being taken as an object in the world, whose agency interacts and affects that which it names. In the case of photography, a social semiotic approach does not need to make a hard difference between visual and linguistic signifying practices; rather, the two are conjoined because seeing, thinking and speaking are continuous aspects both of consciousness and its given world. Philosophy, on the other hand, first and foremost separates itself from the world in order to enquire into the fundamental nature of existence, knowledge and thought, in which language is the means to develop thought. In philosophy, the relationship between photography as a technical image system and philosophy as a linguistic system of thought in language becomes one of philosophic enquiry.

Peter Osborne acknowledges the conjugal relationship between philosophy and photography when he says, 'That is to say, one cannot subtract the historical character of the image from the image of thought' and that 'The attempt to rethink the nature of the photographic image, post-digitalization, has consequences for the reimaging of thought and hence for philosophy itself' (2010, p. 60). In chapter 5 of his book *Anywhere Or Not At All* (2013), titled 'Photographic ontology, infinite exchange', Osborne offers the following account.

> Understood historically, the question of the ontology of the photographic image is in large part the question of the mode of unity of the relational totality of the variety of different photographic forms coexisting within the present: chemical photography, film, television, video and digital imaging – to name only the five main forms – the spine, if you like, of a still expanding field. This totality is relational, rather than expressive, because as a cultural-historical form there is no single underlying ontologically fundamental basis for this unity – in a single technology for example – which would allow for the specification of photography as a medium. (2013, p. 118)

To repeat, for Osborne, photography has no single ontological base as a medium, but rather photography is the mode of unity, the 'idea' of a relational totality whose ontology is only discoverable in/by theory. Here he enters philosophy proper in proposing Kant's idea of a 'distributive unity' as a threat to the establishment of common ideas, which he then connects to its subsequent extraction into 'distributive difference' by Deleuze. Both Kant's 'ideas' and Deleuze's 'difference' are the ontological means of dealing with the threat and philosophic possibility of the infinite multiplication of singularities, which would make the world unintelligible. Osbourne suggests that, 'Such a concept of distributive unity – would articulate the logical form of the historical unity of empirical forms' (2013, p. 122). Osborne's philosophic framework leads him to define the ontology of photography broadly as a pragmatic distributive unity based upon the relations between technologies and the commonality of their functions, a definition which sits happily with a social semiotic account. However, this is not the end of the investigation for philosophy, because the 'commonalty of function' produces a 'de-materialised generality' that transcends material technology, in which ontological meaning of 'the photograph' or the 'still' serves as a metonymic model for photography as a whole. Osborne skirts around giving the founding unity of photography any reductive ontological essence, concluding that the unity of photography is essentially imaginary or mythic, in this case in its reductive identification of cultural form with technology. It is the social actuality of this mythological identification that gives social reality to photography as a cultural form, and hence for Osborne its ontology is to be sought as a cultural category, the unity of which is based on the imagined and practical unification of a particular technological process and a particular set of social functions. One of the conclusions which might be drawn here, as Osborne recognises, is that computational culture presents new challenges to the discipline of philosophy – a challenge to rethink the knowledge boundaries, objects and methods of philosophy, science, political economy and the humanities.

In *Towards a Philosophy of Photography*, Flusser (1983) points out that images are needed to make the world comprehensible and that writing, like images, is a mediation of that world. Flusser establishes an historical dialectic between imagination and conceptualisation, between image and thought. He does this by suggesting that in some earlier pre-given time of 'being,' the circular time and space peculiar to an image belonged to a magical consciousness. Flusser argues that the development of mediating technological apparatuses, of which the first revolution was writing, led to an inversion between world and image, where in interpreting images as texts the world becomes like an image and hence the world is hidden. For Flusser, texts do not signify the world, but instead 'signify the images they tear up' and therefore texts are the metacode of images. With the advent of photography, understood by Flusser as the second revolution of the technical image, the photograph realises the world 'magically' once more, but through the constraints and logic of concepts coded into its apparatuses, thereby further removing a direct conceptualisation of the world.

On many accounts, Flusser's highly insightful analysis of the operations of the photographic apparatus can be made consistent with historical and material accounts. However, Flusser's argument draws upon key concepts from Martin Heidegger's phenomenological philosophy in *Being and Time* (2010 [1927]), which led him to frame photography in terms of an abstract philosophic model of the technical image and its relationship to existence, rather than accounting for the historically specific concrete and political terms of actual socio-technological systems. This is one of the ways in which such abstractions gloss over the radical differences between the photographic image and the computational image. Philosophy renders photography in its own image and in the service of its own ends, although, as illustrated further, not all with the same implications for the relationship between photography and technology. In staying with the attempt to look at how philosophy provides the basis for the slide between an ontology of photography and the technological image it is necessary to consider the impact of Heidegger's philosophy upon both Flusser's and Stiegler's views of technology in more detail.

On the Origins of Technology in Martin Heidegger's *Being and Time*

The resurgence of interest in Martin Heidegger's philosophy and its influence upon thinking about photography is important for a number of reasons, not least because technology has become a new focal point for thinking about culture in general. But interest in phenomenological philosophy also represents a marked turning away from historical and sociological explanations of media, culture and society, which have been seen as inadequate in accounting for subjectivity and experience. Heidegger is a source for those interested in maintaining the unity of the concept of photography, or, as forgetting photography would have it, the refusal to confront the afterlife of photography. Heidegger's philosophy offers an alternative to and a rejection of accounts of photography in terms of the historical development of modes of production, by insisting on the a priori question of the mode of Being in which history is enfolded into the time of Being.

Heidegger introduces a metaphysics which demands that philosophy asks what it means 'to exist', which leads him to develop a method and language to reveal the essence of Being and phenomena encountered in the world. Heidegger argues that the meaning of Being, and hence being as such, had been forgotten by Western philosophy. Heidegger offers the promise that by examining the mode of Being, all essences can be revealed. The starting point of Heidegger's investigation lies in analysing the mode of Being realised by humans, which he terms *Dasein*, the 'there-being' or dwelling in the world, which denotes the entity that human beings as such are. For Heidegger, *Dasein* operates with a pre-ontological understanding of Being which contains a distorted or buried grasp of the a priori conditions underlying the meaning of Being. The task of philosophy, and of what he terms authentic being, is to uncover or reveal the true possibilities of Being. Heidegger laid out the framework for his thinking in *Being and Time* (first published in 1927), in a complex abstract and metaphysical set of terms beyond the remit or grasp of what is at issue here. The point is that the pursuit of photography's essence requires an entry into philosophy, whose extreme logic is photography as philosophy.

But this is not the only path for photography opened by following Heidegger, the other route being photography's relationship to the essence of technology. For Heidegger technology defines the current era, the modern mode of living. Heidegger initially takes technology as tools, from the hammer to complex instruments, that humans have developed to exploit nature. However, Heidegger then says that this instrumental view of technology, as a means to an end, misses the essence of technology, which is a technological mode of Being, a mode of revealing. The modern technological mode of Being corrupts nature through its exploitation. Technology instrumentalises life, a form of not-being, which drives out any sense of the sacred in the presence of beings. But humans are indifferent to this loss, filling it with a technological substitute in the experience of entertainment and information. For photography as for the technological mode it is both of human making and a revealing, an 'unconcealment'[1] of the mode itself. Heidegger argues that the mode of Being needs to undergo a transformation in order for technological devices to be enjoyed and used but not be the overpowering mode of encountering entities. One of the ways in which this can be achieved follows Heidegger's belief that premodern artisanal technologies were in deep harmony with the essence of their materials, a process he termed 'poiesis,' which 'brings forth' a truth of Being-in-the-world.

Heidegger did not write directly about photography, but his thinking about technology as both a forgetting of Being through concealment, but also the possibility of unconcealment as a truth of Being, has been taken up in ontological interest in photography, amongst others by Daniel Rubinstein (2019), in proposing photography as a technology which goes beyond the limits of representation set by language and therefore can be understood as a 'stepping back' from metaphysics and hence a revealing form of thought. From the perspective of forgetting photography, the separation of photography from language is an arbitrary step and a formal abstraction, since the photographic image in use and in the archive is never separate from language in thought and its transmission, an issue considered in more detail in the concluding chapter. But the important issue is that the phenomenological point of view argues that photography as a technical medium conceals the true state of Being, but which, through a philosophical analysis, can reveal the truth of the technical mode of being. The nub of this claim rests upon a test of how photography, or the technological apparatus, becomes unconcealing, how it manifests itself in practical social forms of imaging, or whether it remains a philosophic method. It is upon the question of whether a reformulated version of photography in computational culture can reveal truths of the world that the argument for a new era of photography rests.

Towards a Philosophy of Photography

Flusser's view of the camera as semi-automata was a sign of photography's non-human presence. His argument that photography is a programmed apparatus also makes it other than human. However, there is also a redemptive side to Flusser's argument in which photography is a medium capable of returning the realised human and hence *Towards a Philosophy of Photography* has been taken up as championing the cause of post-photography, by revealing photography's technical essence, through which a new freedom can be won. This is encapsulated in Flusser's view that the photographic universe may serve as a model for post-industrial existence in general and, more pertinently for this discussion, why a philosophy of photography is a point of departure for considering the future form of human existence.

Flusser (1983) used photography as a prime example of a centralising technical information apparatus, which he saw as inevitably leading to automation and along with it a redundancy of human capacity for independent thought. But it would be a mistake to consider photography in its many forms of cultural practice as the real object of Flusser's philosophy of photography; his project was instead a philosophic defence of the freedom of thought from a system of technical apparatuses which were replacing any direct relation humans had with being in the world. Flusser saw in the convergence of media technologies in computing the dominance of a scientific, technical rationalism, which was concealing the existential condition of possibilities in being. His ontological hermeneutic method had allowed him to illuminate through photography how apparatuses had come to dominate, through the distinction he made between historical (natural) and technical images. In his subsequent work Flusser adopted an inclusive view of media as apparatus, focusing more upon the passage from the televisual, rather than photographic image, to the Internet, as did Stiegler.

The technical image is produced by apparatuses, which are themselves the product of scientific texts and therefore are a production of information. Far from seeing photography as a 'natural' invention arising directly from the world, Flusser sees photography as a product constrained by its apparatus. The technical image, which displaces text, removes the necessity of conceptual thinking and replaces historical consciousness with a second-order imagination, the result of which is mass culture. The technical image is the product of a complex system of interrelated apparatuses in which a fundamental inversion has taken place. Whereas in 'traditional' society the tool served human purposes, with industrialisation, this relation was inversed so that humans served and were subordinate to their apparatuses. The future point of the industrialised process is the programmed, post-historical society, no longer based upon causality, but absurd coincidences in which time and space can no longer be considered as separate entities. In this argument Flusser is following a Heideggerian view of technology.

Flusser's view of apparatus follows that of Martin Heidegger, who viewed technology as creating an unfree universe, in which only thought offers the possibility for humans to give significance to their lives in the existential temporal sequence before death. The photographic universe is part of a functionally programmed universe from which individuals need to break free and philosophy offers the possibility of photography becoming a liberating agent. Breaking free of the programmed universe is also an aim of forgetting photography, but not in terms of arguing that philosophy as photography is a means to liberation from the programmed universe. Why photography is given this privileged position within all media apparatuses, which Flusser acknowledges converge in computing, remains a puzzle. In the claim for a liberating future for photography, the technical black box, which encodes the programmed rules of the camera, has to be played against. The basis of Flusser's argument, that 'photographs suppress our critical awareness in order to make us forget the mindless absurdity of the process of functionality, and that it is only thanks to this suppression that functionality is possible at all' (1983, p. 64), is good enough reason to forget photography. Flusser's argument about photographs functioning to conceal functionality can be applied to the social afterlife of photography, which functions to cover for the fact that the apparatus of functionality is no longer photography. The problematic issue in philosophy's discussion of photography is the retention of the historic idea of photography, whereas the actual qualities of photography which philosophy identifies, image-apparatus-programme-information, were never exclusively confined to something named photography, but distributed across apparatuses, and now reside in something other than photography. Flusser would not disagree with this view. Two years after the publication of *Towards a Philosophy of Photography*, Flusser published *Into the Universe of Technical Images* (2011 [1985]), in which the technical image is explicitly a general media category of film, video, computer graphics, holography and virtual reality. Flusser's dialectic is expressed in the following terms: 'Taking contemporary technical images as a starting point, we find two divergent trends. One moves toward a centrally programmed, totalitarian society of image receivers and image administrators, the other toward a dialogic, telematic society of image producers and image collectors. From our standpoint, both these social structures are fantastic, even though the first presents a somewhat negative, the second a positive, utopia' (2011 [1985], p. 4).

This makes attempts to enlist Flusser's earlier analysis of photography as a cornerstone of a specific ontology of photography problematic. Flusser on the other hand was attempting to identify future directions of the informational society and how a new form of freedom based upon what he termed net dialogue might be possible within its emergent dominant medial form. However, Flusser's phenomenological perspective and method has been taken up in post-photographic discourse in arguing that his analysis sheds light upon the apparatus-operator complex, in which each makes the other's existence possible, and each defines the other. What we see in his later writing is that Flusser used photography as the origin of the technical image, in order to make an argument for the transformation of the entire apparatuses of informal functionality that would bring about freedom, and this could only be achieved from within the life world of human experience.

In 1976, Kodak, the iconic giant of the twentieth-century photographic industry, founded by George Eastman in 1888, still sold 90 per cent of all photographic film and 85 per cent of its cameras in the USA and up until 1986 still employed 145,000 people worldwide. In 2012, Kodak filed for bankruptcy. Kodak's demise was not simply a case of not adapting to changing technology; in fact, it was a Kodak engineer, Steve Sassons, who had made one of the earliest digital cameras in 1975. Rather, it was Kodak's addiction to the profit derived from large-scale chemical film and paper sales. Vilém Flusser's book *Towards a Philosophy of Photography* was written over and into a period of the early development and transition to the universal adoption of digital photography, which was eventually integrated into what has become the digital information and telecommunication complex. Digital innovation in the apparatuses of technical reproduction effected not only the capture of images but just as significantly the post-production of image and text, including image transfer, picture editing, design and layout and printing, which developed at roughly the same pace as digital photography. In 1988 the first JPEG standard was set and two years later Adobe Photoshop was launched. The technical means of photo-mechanical reproduction in the print industries gave way to digital reproduction over the course of three decades in a reconfiguration of information channels in which the Internet swept all before it. But while the Sony Corporation demonstrated the Mavica camera which replaced chemical film with magnetic video images in 1981, it was not until well into the 1990s that digital cameras entered the mass market, while news and photo journalist photographers remained sceptical of the quality and permanence of digital imaging and continued to use film for a further decade. Poignantly, many of the professional still images of the destruction of the Twin Towers, leading to the death of 2,977 people on 11 September 2001, were recorded on chemical film. The events of 9/11 marked not only the end of twentieth-century optimism in and for the future, but also a mourning for the loss of continuity with its past. In many ways, 9/11 marked a break with the confident modernity of the twentieth century.

What is important to grasp in Flusser's thinking is that photography is a product of the apparatus upon which it is dependent, producing an entire system based upon the function of the technical image, which is an image not of the phenomenal world but of scientific texts, upon which apparatuses developed. Photography is privileged by Flusser because the camera illustrates his view of the programmatic or cybernetic functions of apparatuses, or, as Osborne (2013) puts it, photography operates mnemonically for the mode, or, more philosophically, the unity, of all apparatuses. What was startling for the photographic community accustomed to understanding the meaning of the photograph to be the result of a photographer's intentions was to hear that the photographer's intention is a function of the camera programme. Flusser defines the act of photography as that of phenomenological doubt because it attempts to approach phenomena from any number of viewpoints, but which are nevertheless prescribed by the camera programme, the apparatus. This results in the photograph being a concept encoded as a state of things, including the photographers' concepts, programmed into the camera, which manipulates us to act in a ritual, programmed fashion in the service of a feedback mechanism for the benefit of the camera (1983, p. 64). Thus, in Flusser's philosophy of photography there is a struggle within photographic practice between the apparatus and phenomenological recognition, which takes the form of a game played against the camera programme. In order not to produce more mindless photographs, of which Flusser sees domestic photography as the prime example, the photographer has to be conscious of the camera programme and calculate images that might reveal something of the way the apparatus interposes itself. The camera provides an accessible point of entry to the apparatus which can be turned against it with the aid of a philosophy which reveals the logic of the black box of the camera programme.

Bernard Stiegler: Technics, Time and the Age of Disruption

In *Technics and Time: The Fault of Epimetheus* (1998), Bernard Stiegler defines technics as the horizon of all possibility to come and of all possibility of a future. Stiegler argues, in the same manner as Heidegger, that philosophy has repressed technics as an object of thought and hence technology is the unthought. Against this Stiegler recognises a new urgency in making the deep opacity of technics and what is being performed and transformed by a technical system of power intelligible, because it is the most powerful and dynamic factor in the numbing of decision making. To paraphrase, Stiegler says that technisation based upon calculation has driven Western knowledge to forgetting its origins and its truths, science having lost the object itself of any science and without a refoundation of rational philosophy will lead to the technisation of the world. As Stiegler says, 'This constitutes a depoliticization of society and promotes a tendency toward the autonomization of purposive rational activities, an evolution that "does harm to language" (Jean-François Lyotard will take up this theme), that is, to socialization, to individuation, and to intersubjectivization' (1998, p. 12).

For Stiegler, as with phenomenology in general, technology cuts both ways, as the thing that enslaves and the thing that has the potential to liberate in grasping its ontology and hence opening the possibilities of becoming fully human. Tools preserve the memory of human gestures which produced them, much as we saw previously in the ideas of Kittler and Ernst in media archaeology, and it is this past recollection in the materiality of technique that constitutes temporality. For Stiegler, photography is a tool specifically dedicated to the support of memory, a specialisation of the memorial dimension of all tools. In *Technics and Time, 2: Disorientation* (2009), Stiegler looks at the development of photographic technology as the industrial production of the past, intensified by measurement and objectivity (2009, p. 42). Observing the world with photographic technology had the effect of readjusting rites and negotiations of experience, memorialisation and reproduction. Photography as an industrial system exteriorises the experience of the viewing subject, making it discrete, and hence is an adjustment of the mediation of Being and becoming. The industrialisation of memory brings with it a new phase in the grammatisation of experience, recollection and consciousness (2009, p. 41). The salient point to comment upon here is that Stiegler maintains a distinction between photography and the digital convergence of telecommunications and computing, which he sees as centred upon what has happened to television. Photography in Stiegler's treatment remains with its analogic roots and earlier industrial capital. Stiegler points out the even greater integration of analogue communication industries, journalism and editorial functions that have taken place over the last two decades, leading to the digital information industries and what he calls a reticular society based upon a data economy which systematically exploits interindividual and transindiviaul relations in a form of algorithmic governmentality.

For Stiegler (2016) the relationship between photography and technology takes a decisive shift away from the discourse of photography towards a full embrace of what he terms 'the digital,' which marks an immense transformation exploding all frames of thought, 'something other than the computerization of society: it was an explosive process, a combinatory explosion that provoked a chain reaction we now call disruptive, destroying the frameworks of nearly every domain and questioning, to an unthinkable degree, public power' (2016, p. 159). Stiegler is giving a new worldly voice to philosophy by translating the logic of abstract technology and its phenomenological conception into a political programme. The dialectic between the toxicity of technology and its potential for human liberation is still evident. He says, 'The digital has a fundamental positivity in that it is the vehicle of a process of deproletarianization, a reconstruction of processes of psychic and collective individuation, going beyond the opposition between production and consumption and, with that, the industrial division of labor described by Adam Smith.' This is an altogether different kind of philosophical engagement with the world and, in the same manner as Osborne's comment about the effect of the new technological conditions upon philosophy, Stiegler recognises the radical consequences of the digital on knowledge paradigms.

Photography and the Computational Image Part Company

In *Signal, Image, Architecture,* John May (2019) considers the status of computational images in contemporary architectural thought and practice by showing what happens if the technical basis of architecture is closely scrutinised. Drawing upon philosophic concepts, May lays out three axioms from phenomenological philosophy in order to put forward the view that the computational image has nothing in common with either the photographic or orthographic and that the idea of the digital photograph is an oxymoron. May's primary purpose is located in architectural education and practice and his analysis questions the consequences for the architectural imagination of using computational imaging software. In doing this he is necessarily drawn to account for the photo-realist basis of graphic imaging and hence considers photography.

May outlines three axioms which guide his analysis and which are drawn from the philosophic discourse on technology derived from Heidegger and Steigler, which state: 1) there are no pre-technical forms of thought; 2) all technicalities are tethered, in some way or another, to the deepest regions of consciousness; 3) the specific conception of time embedded in a technical system is inseparable from the forms of thought and imagination it makes possible or impossible. Using ideas strikingly similar to Heidegger's concept of 'Being' and time, May is able to question the use of the digital and post-digital as terms of linear historical development. As May sees it, contemporary architecture seems uninterested in distinguishing between drawings, photographs and images, and:

> Far from indicating the permanence, exchangeability, or 'resurgence' of any one category, the slippages between them is symptomatic of a chronic confusion, throughout the design fields, that is rooted in basic category errors concerning the enmeshment of life, thought, and technique. (2019, p. 11)

By seeing technology as inseparable from consciousness and time May is able to apprehend photography phenomenologically, as others have done, as light writing, a chemo-mechanical inscription of an instantaneous scene, much as Barthes describes the perfect analogon. May argues that from the perspective of visual representation the invention of photography amounted to a regression of 'visual mathematiziation'. This is because the mathematics of a photograph always remains locked deep within its chemistry, offering no empirical surface for immediate calculation. May goes on to argue that from the interests of calculation, the impenetrable analogue of the photograph was the stimulus for the development of the electrical image, amenable to calculations and derived from technical domains unrelated to the practice of photography. The lineage of the computational image is electrical engineering, telegraphy and physiological optics, which leads May to conclude that images are a form of photon detection whose outputs are defined by signalisation, which accumulatively amounts to data and its processing.

For photographic discourse and its attempts to bridge the digital divide, the weight of May's argument is devastating.

> Photography and photo-detection are held apart from one another as technical categories by this fundamental and unbridgeable epistemic abyss between heliography and bolometry, between photography as written light and imaging as detected energy, between Nicéphore Niépce's View from the Window at Le Gras (ca. 1826) and John Logie Baird's telescan of Oliver Hutchinson. (2019, p. 13)

May is able to conclude that photographs and computer images have virtually nothing in common with one another, other than a misleading visual resemblance that equates two completely incompatible technical formats. For May, the technical format of the photograph and that of the photon-image belong to two competing epistemic visions of the world and those images should be regarded as manifestations of spreadsheets and statistical formulas which are structurally calculable, and only apparently visual. Here, it is possible to see that a phenomenological view of technology developed by an interest in the ontology of graphic software ultimately distinguishes photography from the screen's image. May provides a compelling argument for forgetting photography and seeing the contradications held by the terms digital photography and computer graphics, which wage a futile war against technical obscelence. May concludes, 'It is a campaign waged entirely within the terrain of the psyche, where our language lingers stubbornly, and where concepts that long ago lost their purchase on reality persist as nothing more than collective mental habits' (2019, pp. 13–14).

Philosophy Becomes Photography

It might be considered that François Laruelle in his book *The Concept of Non-Photography* (2011) is putting forward an argument sympathetic to photography's afterlife, and indeed by conceptualising the foundation of the world and philosophy with it as essentially photographic, he places photography outside of itself, arguing that if the world is photographic, how could it be photographed? For Laruelle photography as an image technology has to be regarded as a phenomenon of the pre-given a priori photographic world and hence provides a 'clone' or a fiction of the world. Laruelle developed non-photography as a departure from what he calls standard philosophy, which presupposes that it is epistemologically sufficient to decide on the nature of the real. But for Laruelle the 'Real' cannot be encompassed by philosophy because the immanent totality of the real, understood as all that is, has to exceed any conceptual decision that philosophy defines. In developing his concept of non-photography, Laruelle argues that the 'All,' of 'all there is' would have begun with a flash.

> Such is the philosophical legend of the originary flash, of the birth of the World, a legend of the birth of philosophy in the spirit of photography. Philosophy announces that the Cosmos is a 'shot,' and announces itself as this creative shot of the World. Heraclitus' child at play would, in the end, have been nothing but a photographer. (2011, p. 1)

From this point of origin Laruelle argues that there is no point in trying to separate philosophy from this photographic legend that encompasses it, because even before its invention as a technology Western thought was suffused by photographic repetition. Thus Laruelle is able to pursue an ontological elaboration of philosophy as a photographic catastrophe, 'as an irruption of the "empty" essence of photography and as an intoxication of All-photography and of the photography of the All' (2011, p. 3). His term non-photography describes a new conception of the essence of photography, the practices which arise from it and of its relation to philosophy. There is no longer a need to think photography through philosophy, but instead to seek 'an absolute non-onto-photo-logical thinking of essence' (2011, p. 4) in order to understand what photography is and what it can do. What Laruelle calls 'photographism' takes the place of thought and its effects constitute a form of forgetting of the essence of photography. In this conclusion, photography cannot be historically forgotten, because it has been given an atemporal and apriori universal condition which the efforts of philosophy, now as photography, can reveal.

Conclusion

The problem identified here is that, in the strict sense of the designation, photography in philosophy is as an abstraction, which in seeking the essence of photography, either in its apparatus or image, operates on a plane necessarily separated from that of contingent social encounter.

Idealist philosophy is rarely willing or able to grapple with concrete historical circumstances and changing material conditions. Because of the philosophic default to abstraction and the essence of things, the use of phenomenology in the discipline of philosophy – what might now be termed traditional or academic philosophy, rather than worldly and engaged epistemologies – runs the risk of misrecognising that the current socio-technical basis of the network image has left photography behind. The picture of the discourse of photography which bends towards the philosophy of photography is marked by a lack of resolve and a means of returning the discourse to the concrete social world and lacks a commitment to a practical politics, even though it is a philosophy of the existential condition.[2] What does emerge is a perennial slippage between the photographic and the digital, which are not at all the same thing, and at a time when theories of visuality, meaning and value are moving and expanding to keep pace with the global, computational and networked image, the retention of the philosophic discourse of photography can only be a conservative falling back and a reason why the much-needed enterprise of clearly articulating and expressing a common and hence political view of the contemporary condition of visuality and mediality in culture is proving so hard to achieve.

Photographic theory appears as the scaffolding of photography's past but remains present as its abiding architecture. From the position of photography's afterlife, understanding photographic theory becomes a work of genealogy. Under the sign of photography lie countless hybrid practices, whilst theorising image hybridity (otherwise known as photography) has historically been the opposite impulse of distilling its singularity. The preoccupation of photographic theory reflects not so much the dispersed and divergent practices of photography in its exponential expansion in everyday life as it does the epistemological interests of the discipline from which theory is enunciated. There is a self-reflexive fascination in observing that one of the central strands of photography is the theoretical determination to define its essence. From such an observation it is possible to discern a descending, or possibly competing, order of theoretical frames of reference, which attempt to capture and define an abstract singularity of the object of photography. In attempting to characterise some of the ways in which phenomenology has been taken up as a means to define the essence of photography, it is immediately evident that it is subsumed in a more all-encompassing abstract category of technology, which presents a problem to the historical and material argument upon which forgetting photography rests. The argument made here is that the ontology of photography is its remembrance.

5

Photography and Modernism: A Case Study of Tate Modern and Tate Britain

These images immediately bear the stamp of the laws of knowledge. The indeterminate, since it does not allow for precision, will have to be eliminated, and with it goes feeling. The person for whom these beautiful pictures are intended is a consumer of finished products. Photography's infallibility is that of the perfectly programmed; its beauty is that of Voyager II.

Jean-François Lyotard. 1982. 'Presenting the Unpresentable: The Sublime'. *Artforum*, 20(8), p. 65.

Post-photographic discourse perpetuates the jumbling of photography with technology, photography with art and photography with the contemporary, a mix which is confusing and needs sorting out, which is what the strategy of forgetting photography is dedicated to. To move beyond or outside of photography, it is necessary to remember what photography as a medium and a culture has been and what it has done, rather than what it still purports to do. Put another way, this is the effort to see photography as an historically bounded assemblage of ideas, objects and events. It is through the remembrance of photography and identifying its memorialisation that photography can ultimately be forgotten and in doing so make the clearing for a new understanding and language of the image and representation, as well as seeing the world photography showed. Inevitably the remembrance of photography involves, by definition, nostalgia, mourning and a deep sense of loss, which can be found in every exhibition of photography. There is no way around photography unless the zombie manifestations of photography are identified. Forgetting photography involves its memory. The afterlife of photography is manifest in contemporary exhibition and this chapter takes Tate Britain and Tate Modern as a case study. Forgetting might seem disingenuous by now, given the privilege afforded to photography's continued presence. In mitigation, given the argument that photography has died several deaths, the method of forgetting photography involves its remembrance, even though, paradoxically, photography is still everywhere.

In November 2014, Tate released a press statement announcing its 'continuing commitment to photography'.[1] Like a guilty secret, the phrase introduces a note of doubt on the very thing it claims to have, a commitment to photography, as if Tate knows there is a whispering campaign which says, 'Tate has never been committed to photography'. But what is a commitment to photography anyway, approaching two centuries after its arrival and in its current post-condition? What form might a commitment to photography take in the art museum? Tate's policies, collection, exhibition and display practices are a means of understanding photography's precipitous relationship to art, to the formation of a canon and to modernist formalism.

This chapter is an attempt to show in detail how getting beyond photography is an institutional and temporal problem. It also reflects a familiar and fond relationship with Tate dating back to the 1960s, when what is now Tate Britain at Millbank was visited mostly by a few art students, stray tourists and 'ladies that lunch' (this was a phrase used by curators as an internal shorthand, a kindly joke, to describe a perennial section of Tate Britain's core weekday audience). If photography was present at all at Tate in the 1960s, it was rarely displayed, and when it was, it was because it was used as a source or medium in a work of art.[2] Otherwise, it was assigned to the archive, not the collection. This is hard to imagine now with the phenomenal success of Tate Modern, opened in 2000 in the former Bankside Power Station, redesigned by Jacques Hertzog and Pierre de Meuron, and whose visitor numbers have grown annually to over 6 million. Tate Modern, under the directorship of Sir Nicholas Serota, established an art museum brand that might rival Nike, banishing forever the image of the museum as a preserve of a cultural elite, and setting a precedent for the new corporate global art museum, a new kind of heterotopia with a revolving door to admit post-traditional audiences to new forms of the consumption of art.

Tate is responsible for the National Collection of British Art, dating from the 1500s and comprising nearly 70,000 works. It has an annual display and exhibition programme across its four galleries, Tate St Ives, Tate

Liverpool, Tate Britain and Tate Modern, showcasing the work of British and international artists. Tate's policy on collecting British Art emphasises that artists are chosen for their contribution to Britain's history and development, rather than their nationality alone. The gallery was first opened to the public in 1897, built on the site of a former prison, the Millbank Penitentiary, used as a departure point for convicts bound for Australia. Tate was under the directorship of the National Gallery at Trafalgar Square until 1955, when it became wholly independent. The original bequest of works by Henry Tate, who made his fortune in refining imported sugar and whose name the gallery adopted, has led to longstanding speculation about his relationship to slavery, about which Tate has for some time been concerned, recently collaborating with the Centre for the Study of the Legacies of British Slave-ownership at University College London on researching the connection. In 2019, the UCL project concluded that Henry Tate had no direct connections with slave ownership, but that it is also 'not possible to separate the Tate galleries from the history of colonial slavery from which in part they derive their existence'.[3] Such a statement might also be made for photography, as discussed in relation to Jonathan Beller's book *The Message is Murder: Substrates of Computational Capital* (2018) in Chapter 2. Educational and public interest and concern over the colonial histories of museum collections is a current manifestation of the crisis of representation. Questioning the provenance of museum objects and the sources of wealth from which collections were made comes at a particular conjuncture and is a politics on several connected levels. Clearly, the crisis of representation and the gatherings around identity politics are enmeshed in the stalling of democratic reform and progress, by states not prepared or able to regulate neoliberal capital accumulation. The historical and systemic force of racism, which Beller sees as endemic to photographic representation, provides a political analysis and an educational agenda for museums, public education as well as a further way of seeing the history of photography.

Tate's commitment to photography came late in the day, with the appointment of its first dedicated photography curator Simon Baker in 2009 and the formation of a photography acquisitions committee a year later. Tate's late adoption of photography stands in marked contrast to that of the Museum of Modern Art (MoMA) in New York, whose first photography survey show, *Photography: 1839–1937*, was curated by Beaumont Newhall in 1937, and he established a photography department in 1940, creating a separate centre dedicated to photography to provide archive and exhibition space to its collection of over 2,000 prints. The adoption of photography in the early decades of the twentieth century was expressive of the foundation of the USA as a nation, its expanding economy, immigrant culture, anti-communist paranoia and capitalist imperialism. Photography was a medium suited to traversing the continent's large and diverse geographies and exporting the values of American culture abroad, well reflected in the exhibition *The Family of Man*, curated by Edward Steichen, photographer and director of MoMA's photography department in 1955 at the height of the Cold War, which travelled to 37 countries over six continents for eight years. Modern medium, modern America, but not so in the 'old' country, where the legitimation of photography took a different and slower path, remaining an adjunct to fine art establishments and one of the reasons why the historical canon of international photography remains centred on the USA.

It is widely accepted, including by Tate, that its first major survey exhibition dedicated to the medium of photography was *Cruel and Tender: The Real in Twentieth-Century Photography* in 2003, curated by Emma Dexter, a senior curator at Tate, and Thomas Weski, chief curator at Museum Ludwig. This was a major collaborative exhibition of over 600 works, organised around a number of social themes, including occupied spaces, vulnerability, life stories, industrialisation and consumerism and 'on the road.' Interestingly the themes mirror those of established histories of photography discussed earlier, in the ways in which categories of the real and the photographic real seamlessly merge and overlap. The exhibition was not hung chronologically, beginning and ending with the work of living photographers, as if to attest to the contemporaneity and modernity of the documentary form. The art critic Jonathan Jones, complaining about the exhibition at the time, expressed the view that putting photographs in frames on gallery walls is the opposite of the way we experience photographs in real life, and that photography cannot rival the visual and intellectual fullness of high modernist painting and sculpture. Jones viewed modernist photography as 'silly' and at best a 'curio,' and photography in general at best as information.[4] This defence of modernism in high art, made by asserting the lack of distinction of photography, misses the wider picture, which was that well before 2003 the discourse of modernism was an unravelling historical category. But even in the face of major movements of the tectonic plates of Western intellectual thought, art and photography have continued to be defended by modernism's claim to the contemporary and to speak for the present. Tate had made the distinction between art photography and conceptual art in the 1980s, when it added photographic works by artists but not 'photographers' to its collection. The fact that the distinction between art and photography at Tate subsequently melted away was not simply because photography had finally been accepted by the art institution and the art market. The acceptance of photography was a consequence of a deeper reconfiguration of global economies accompanied by an undoing of modernist intellectual certainty, upon which the distinction between art and photography rested and which had kept photography out of the art museum. Indeed, the period saw a boom in the global art market in which photography became an investment like any other art commodity. In 2007, a single print, '99 Cent II Diptychon' by Andreas Gursky, sold for £1.7 million at Sotheby's London. The unsettling of modernity being referred to is not simply to be registered in art historical and curatorial terms, but in the much more real and radical terms of the culture of everyday life. The early success of Tate Modern embodied the new terms of the society of the spectacle, of culture as event, of art as a form of instantly accessible commodity and artists as celebrities, all qualities which emphasised the new mode of the attention economy, itself the result of commodification and information.

The unsettling of modernity, the admittance of multiple modernities, with differing locations and temporalities, which Tate Modern embraced while continuing to curate under the auspices of modernity is one of the striking paradoxes explored in what follows. By the same token the increasing inability of the art museum to legislate the constitution of the modern opens up different histories and perspectives on twentieth-century photographic practices, so far situated on the margins of interest to museum collections. The aim of excavating marginal, forgotten and unrecognised photographic practices is not undertaken to gain their belated admittance to the canon, but rather to question the canon and build new political micro histories of photography in its aftermath.

A Very British Context: A Micro History of Photography's Afterlife

The admittance of photography at Tate was, in Tate's own terms, its acceptance as a contemporary medium of art; as Simon Baker said, 'We try to keep the photography displays integrated with all the other media, but also keep our ideas integrated. I'm always working on a broader context, which is that we are a contemporary art gallery.' Tate explained the quadrupling of its holdings of photography as 'a reflection of photography's ever more important role in contemporary artistic practices and international visual culture.'[5]

The question as to why Tate and the contemporary British art establishment annexed photography for most of the twentieth century is answered by Jones's prejudice of photography's lack of distinction, precisely in the technical sense that Pierre Bourdieu (2010) designated the term. For Bourdieu, a knowledge and appreciation of art was a means by which an educated middle class acquired cultural capital and hence distinguished themselves. Photography was a wavering category defined as a middlebrow art form in Bourdeiu's class cultural schema.[6] But, in the two decades leading up to the millennium and the opening of Tate Modern, Britain had transformed into a post-industrial society, which fundamentally disrupted the cultural patterns of the particular industrial class society Bourdieu had analysed and upon which distinction operated. Britain's harsh deindustrialisation was presided over by a Conservative government led by Margaret Thatcher, who adopted monetarist policy, reduced public spending, enacted anti-union legislation and inaugurated privatisation, which has remained the basis of Conservative policy since. When, in 1987, Thatcher gave an interview on BBC Radio 4's 'Women's Hour' in which she made her notorious claim that there is no such thing as society, she wasn't being a sociologist but a politician, who couched the dismantling of the postwar welfare state in favour of deregulated markets in ideological and divisive terms, of the interests of individuals and families against the undeserving poor.

In 2008, Thatcher's claim became the title of an historical documentary photography exhibition, *No Such Thing as Society: Photography in Britain 1968–1987*, drawn from works in the Arts Council and British Council collections, curated by David Mellor. Much of the work in the exhibition was collected under the auspices of Barry Lane, the first and only photography officer at the Arts Council of Great Britain between 1973 and 1993, and Brett Rogers, who worked for the Arts Department of the British Council between 1982 and 2005, as deputy director and head of exhibitions. Rogers became director of the Photographers' Gallery in 2005. Photography in Britain, under the odd title 'independent photography',[7] delineated a category of documentary photography distinct from domestic, amateur, commercial and industrial photography. Independent photography was considered distinct from the way photography was used in contemporary art and was championed and supported by the Arts Council of Great Britain through a photography committee established by Barry Lane. Lane built up considerable influence within visual arts at the Arts Council, with an increased annual budget to support independent photographers and award grants to independent photography and galleries. The Photographers' Gallery and Impressions Gallery in York, directed by Paul Wombell, amongst others, were able with Arts Council support to commission touring exhibitions and subsidise photographic publishing. Lane organised the Diane Arbus exhibition at the Hayward Gallery in 1973, *Pictorial Photography in Britain 1900–1920*, and *Neue Sachlichkeit: New Realism in German Photography of the 20s* in 1978. He also commissioned a number of policy reports, including the first national policy for 'Photography and Education' (1987), 'Photography, the Arts and Culture Industries for the National Arts and Media Strategy' (1992), 'Photography/New Media Centres' (1994) and 'Creating Vision – Photography and the National Curriculum' (1994). In 1998, the Arts Council awarded the 'Year of Photography and Electronic Image' to Yorkshire and Humberside, which was conceived and organised by Lane.

This parallel and little-known history of British independent photography is an important piece of the puzzle of Tate's resistance to photography between the 1970s and 2003. Tate's resistance was premised not only upon photography's middlebrow status, but was also a resistance to what independent photography was depicting of British class society in the world outside of the art museum. British independent photography was forged by the consequences of deindustrialisation and the callous support of a Conservative-led state, which was resisted by communities and trade unions and led to social strife and displacement. This was the context in which renewed social documentary and community photographic practices emerged as a response to government cuts, which were disdained by the British art establishment. Barry Lane left the Arts Council in 1995 as a consequence of its decision to dissolve the photography committee, cutting its budget for visual arts on the very argument that there was no longer any distinction between photography and art. Lane went on to direct the Royal Photographic Society (RPS) collection until 2002, when, at a point of financial instability, the collection went to the National Media Museum in Bradford. In 2016, the National Science Museums Group agreed to transfer the RPS collection to the Victoria and Albert Museum, a topic which is discussed in the following chapter.

The belated admittance of photography by Tate came too late for photography as a contemporary medium and in subsequent exhibition and display practice its was photography's afterlife that had been admitted, rather than the default image of visual representation. The commitment and desire of Tate to be at the centre of global visual culture, in which photography was now accepted as a contemporary art medium, was flawed by the deeper undoing of the singular temporal logic of the contemporary[8] and by the new conditions of reproduction in which the temporality of the image no longer denoted a singular present. Tate had admitted not photography as a medium of the present, but as a medium of the archival past in which photography now belonged to a commodified heritage culture. Tate had, not for the first time, played a safe conservative curatorial hand by admitting not the new mode of the image, but its deceased analogue predecessor. More bizarrely, in the moment of admitting photography Tate unwittingly set up a new exclusion zone in order to keep the new ubiquitous, profane hybrid image and its prosumer audience out of the museum and in so doing preserved the category of distinction after all. This can be expressed as the art museum's fear of the Internet.

In the same year as the first major photography survey exhibition, *Cruel and Tender* at Tate in 2003, Myspace, a social media site, was nearing 1 million active users per month in comparison to the respectable 100,000 visitors to the Tate exhibition. The comparison is one of disparate scales of circulation and user participation, evident now in Facebook's membership, which currently stands at 1.69 billion regular users, and Google, started in 1998, which is now processing a staggering 3.5 billion searches per day. In 2003, the Internet was on its second if not third iteration of the Web, which by then meant, amongst other things, that embedded digital media was at the front end of the user interface. That year a number of key new media platforms and applications appeared. Apple launched the iTunes store and Second Life and Skype were released, along with WordPress and Safari. Over the following two years the now familiar landscape of the windowed computer interface to the Web was well established. Facebook and Flickr were released in 2004, YouTube in 2005, Twitter in 2006 and, crucially, Apple's first smart phone in 2007, followed by Instagram in 2010 and so it went on. The Internet network and the protocols of the World Wide Web now constitute a computational megastructure straddling the planet. It enables state intelligence agencies and global data gathering corporations to store and transmit the world's information and to track, survey and harvest personal data. It is the new global social medium of business, intelligence and telecommunications. Why then, in wanting to be at the forefront of contemporary culture, has the art establishment been so reluctant to engage with the most powerful contemporary culture in existence?

The Art Museum and the Disavowal of the Digital

Claire Bishop, writing in *Artforum* in 2012,[9] aptly put the paradox of the digital in the art museum, in arguing that mainstream contemporary art disavows digital culture but depends upon it for its production. She asks why contemporary art doesn't embrace the experience of digital life, whereas in comparison twentieth-century art had a fascination with the analogue formation of the image and its cultural forms of reproduction, particularly reflected by television as a cultural logic from the 1960s. Her answer is that film, photography and video were image-based and hence their challenge and interest to visual art was self-evident, whereas the digital is based upon mathematical and linguistic code, which remains indecipherable in its infinite multiplicity. Since the 1990s, contemporary art museums have installed works by artists displaying a fascination with analogue archives and mediums. Whirring mechanical film projectors, clunking carousels, cathode-ray tube television monitors and vitrines full of analogue print have been the stock in trade of mainstream contemporary art exhibitions. But, as Clare Bishop has observed, apart from a few purists working with the analogue craft of photography and film, much of this analogue media-based work relies upon digital forms of research and production through Internet archival searching, digital post-production software treatment of analogue material, as well as the use of Wi-Fi and digital projection for installations. Bishop argues that behind such work, the 'digital' as she puts it, is the deeper shaping condition that determines artistic decisions to work with particular formats and media, but that its presence remains subterranean and that this can be considered a form of disavowal.

Has the situation changed much in public exhibition since Bishop's article in *Artforum*? The Barbican in London collaborated with Google in 2014 to produce the exhibition *Digital Revolution*, which focused upon technological innovations of augmented reality, artificial intelligence, wearable technologies and 3D printing. The Whitney programmed *Rules, Codes and Choreographies in Art, 1965–2018* in 2018, displaying works from its collection on the theme of instruction-based works of conceptual video and computational art. MoMA curated a survey show from its own collection, entitled *New Order: Art and Technology in the 21st Century* in 2019 showcasing a diverse range of techniques and media, digital simulation, including industrial vacuum-formed plastic and ultrasound gel. In both exhibitions the emphasis was upon innovative industrial technologies, rather than the reorganisation the Internet was establishing in economic, social and political life. Omar Kholeif curated *Electronic Superhighway (2016–1966)* for the Whitechapel in a survey exhibition of over 100 works designed to show the impact of computer and Internet technologies upon the arts, essentially creating a canon for a distinct art genre, harking back to the exhibition *Cybernetic Serendipity*.[10] Since 2015 the Serpentine Galleries have had a dedicated arts technologies strand of programming, including exhibitions by Hito Steyrel and Ian Cheng, Jacob Kudsk Steensen and Suzanne Triester and an R&D platform for Future Art Ecosystems run by Ben Vickers and Kay Watson. The R&D platform takes up the relationship between art and technology in a much more far-reaching programme of thinking about the future organisation of art in the face of the Internet. More specific to photography, the Photographers' Gallery in London has run a Digital Programme, initially organised around a dedicated platform for screen media since 2012, which is committed to making the politics of software visible, discussed in more detail in Chapter 8.

In the wider scheme of exhibitions in major contemporary art museums and galleries exhibitions focused upon technoculture have been a peripheral strand of programming, neither matching nor doing justice to the dominance of the Internet in everyday life and the prevalence of digital technology in art and media production. The criticism still stands that the conditions of the digital mode of reproduction, which has had such a marked effect in everyday life over the past two decades, if not longer, remain invisible in the art museum. The admittance of photography does nothing to redress this situation; indeed, it obscures temporal disjunctures and the nonrepresentational conditions of the computational distributed image. What the disavowal of the digital does do for art and photography is open up a deep well of nostalgia and memorialisation of the analogue itself, another compelling reason to see that photography in the art museum is essentially photography's innocent past life.

The disavowal of the digital condition in the art museum is not confined to what is exhibited and collected, but encompasses the organisational functions of the art museum as well, which continues to struggle with a sustained engagement with the values and communities of the Internet. Tate, in line with the museum sector in general, has used the Internet primarily, if not wholly, as a marketing tool, as an extension of Tate's previous analogue communication department organised around press and publicity.[11] Such departmental organisation is still based upon an analogue model of public-service broadcasting, although latterly taking on the mantel of commodity advertising, in which cultural authority emanates from a central source, in an echo of the few talking to the many. The use of the Internet as a marketing tool and more recently as a media content platform by Tate is not aligned to its desire to realise its goal of universal engagement and inclusion. Even the global pandemic sparked by Covid-19, which closed museums and forced working online and from home for most of 2020, only increased the volume rather than the type of Tate's online activity. And this is because of the adherence to the cultural values of modernism, which still require the art museum to apply the filter of the authority of the artist, the expertise of the curator and the discourse of the contemporary. For Tate to engage with the culture of the Internet would involve an acceptance of the multiplicity of hybrids across the networks, as met in their many different contexts of user engagement on networked screens. To engage with the culture of the Internet the art museum would essentially have to distribute itself, requiring changes to its organisation, operations, skill and knowledge base and new forms of collaboration. The acceptance of photography as a fully fledged art medium, admitted to the art museum, came at the historical moment of its obsolescence, in which its hold over reality had diminished. What then could be the function of the exhibition and collection of photography in the art museum other than the only task left to it, that of weaving photography into the canon of aesthetic modernism, as well as providing its corollary, in illustrating the historical figures and events of modernist art, a fusing together of the history of photography and the historical period of photography? In order to do this the curatorial imperative of the modern in the art museum was the ceaseless work of purification against the hybrids of all kinds, which until recently would have placed photography in the archives.

The Modern

It is worth looking at how Bruno Latour describes the paradox of the modern, as it informs the view of the art museum taken here and runs throughout the overall argument of forgetting photography. The paradox of the modern has a special resonance for photography, given its place between nature and culture as a widely accepted medium of record and time. Photography as a modern medium in Walter Benjamin's sense rests precisely upon a medium matched to its progressive time, the time of the modern. This remains one of photography's central claims to be relevant as a medium for penetrating reality. It is, therefore, not surprising in making the argument that the photography of the modern no longer exists, to find the companion argument that we were never modern.

Bruno Latour's epistemological undoing of the philosophical discourse of scientific rationalism is highly pertinent to the predicament of exhibiting photography. In his book *We Have Never Been Modern* (1993), Latour identifies that the term 'modern' is agonistic and conflict generating, in being both a term of temporal ontology and a critical term. Latour goes on to argue that the tension in the term is evident in the opposition between the ancients and the moderns, through which the term 'modern' first acquired a periodising significance and which by the end of the eighteenth century had acquired the temporal emphasis upon the new. This new sense of the modern registered a break, not merely with the Ancients, but with the temporality of the tradition itself, giving rise to the term 'modernity' and 'modernism' which intensified further the sense of the present as modern (1993, p. 73).

Latour then sets out his hypothesis of the work and practices of the modern, in which he says that the word modern 'designates two set of entirely different practices, which must remain distinct if they are to remain effective, but have recently been confused' (1993, p. 10). Latour terms the first set as practices through translation, which create hybrids of nature and culture, the second set, by purification, create two distinct ontological zones, between humans and non-human. The paradox of the modern is that without the hybrids of nature and culture the practices of purification would be pointless, and without purification the work of translation would by limited or even ruled out. Latour's hypothesis of the paradox of the modern defines the hybrids of nature and culture as networks and the ontological separation of human and non-human worlds as the modern critical stance.

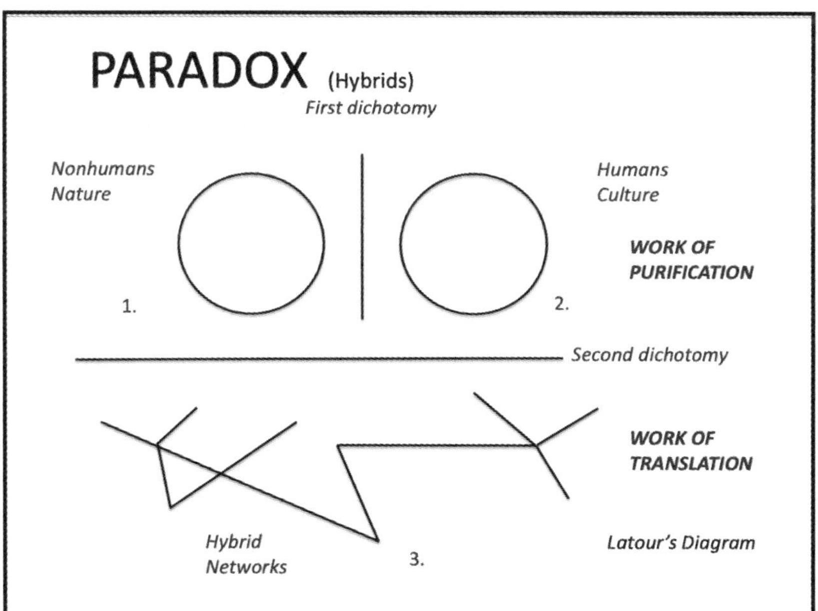

Fig. 5.1

Bruno Latour, *The Dichotomies of the Modern.*

Over the past two decades, academic photographic studies continues to be based upon a modernist purification project against the proliferation of hybrids. The hybrids of nature and culture are constantly multiplying and, in the case of the visual image, dissolving medium specificity into a general state of media convergence in computation and graphics. Meanwhile, the purification project for photography works away at stripping back the photographic image to its singularity. As already discussed, the evidence for this latter-day revisionist project of modernist purification can be found in various forms, in the revived interest in photography's indexicality, its technical materiality and here in collection and exhibition. All of these interests are manifestations of photography's reclamation in a computational world where, even though photography as a specific singular medium is gone for good, it refuses to go. Latour sums up the predicament in the following terms. The modern is upheld as long as the practices of translation and purification are considered separately, and the critical project prevails even though that project is developed only through the proliferation of hybrids. However, when attention is directed simultaneously to the work of purification and hybridisation, 'we immediately stop being wholly modern ... At the same time we stop having been modern, because we become retrospectively aware that the two sets of practices have always already been at work in the historical period that is ending' (Latour 1993, pp. 10–11). Latour characterises the latter retrospective attitude as 'nonmodern' or 'amodern'. Forgetting photography is one such retrospective attitude.

Aesthetic Modernism

The established and normative notion of spectatorship in the art museum entailed in aesthetic modernism is based upon the idea of a singular and exceptional encounter and transaction between a unique viewing subject and a unique work of art, which takes place ideally in the 'purified', apparently neutral space of the gallery. The foundational modernist subject and her gaze is at the heart of the art encounter, just as the work of art is the product of the unique vision of the artist. The modernist gallery viewing space, the white cube, establishes a barrier to the interference, the noise and clamour of everyday life, in order to privilege the possibility of aesthetic experience. The language of visual aesthetic experience triangulates the object seen and contemplated, the abstract concepts of beauty, truth and the sublime and the formal conventions deployed by the object – in other words its artifice and rhetoric. Aesthetic formalism is derived from the study of works of art and their appreciation in terms of the compositional properties such as framing, symmetry, balance, tonality, shape and figuration. In art history, modernist aesthetics is characterised by its divergence from and playfulness towards the aesthetic rules of classicism and neo-classicism. Modernism in art is thus defined by its experimental nature in relationship to the historical project of art itself, hence the movement towards abstraction and conceptualism. Now, in several further moves on from post-conceptualism, contemporary art practice is characterised by the paradoxical present;[12] it still wants to make objects of beauty, which become debased exchange commodities; it rejects the material object but is dependent upon materiality; it wants to be socially and progressively useful, but is constrained by arts institutions; it wants art to be collective, but is premised on the language of individualism.

How does modernist aesthetic formalism in the paradoxical present play out in the afterlife of photography as a practice distinct from and yet taken as art? The common and popular terms for making or encountering a photographic image remain the judgement of what constitutes a good or a bad photograph in formal terms. Outmoded as aesthetic formalism may now seem in the face of the screen image, it still informs the reproduction of an international canon of photography and its collection and display, the marketing of photographic equipment, the aspirations of specialist photographic cultures and the teaching of photography. However, aesthetic formalism in photography comes up against two major obstacles. First, the photographic image is always a representation of something external. However much modernist photographic prints are formally abstracted and organised on a two-dimensional surface, the contingent cultural connotation of the referent imposes itself. Second, the photographic image is part of everyday life, as a commonly met object of reproduction with little aesthetic interest. These objections may have kept the reproductive function of the analogue photographic image out of the canon, but now the networked image is everywhere, in multiple temporal contexts, as a hybridised image, making the purification of photography, necessary for aesthetic formalism in the art museum, a near impossible task. To see photography in the contemporary art museum is to see photography as it was rather than is. In the art museum photography is met as the present that once was. There can be nothing contemporary about the exhibition of photography on gallery walls, because the contemporary image, such as the term still holds meaning, is met on screens as a fleeting absent/presence. But it is aesthetic modernism which conceals the absence of contemporality in photography through its exclusion of the temporalities of the Internet and the trick of reconstituting the linear historical time upon which the contemporary rests. In contrast the screen and network are busily undoing singularity, creating image flows, multiple identities and temporalities in an unnamed space of viewing and being viewed, which challenges the foundational claim of the modernist subject and establishes hybrid aesthetics.

An Ancient Tomb

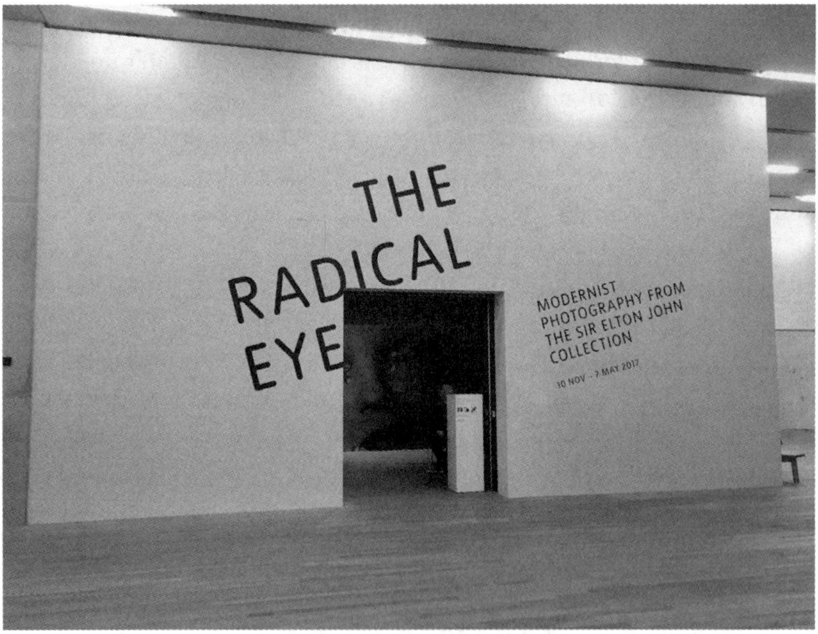

Fig. 5.2

The Radical Eye: Modernist Photography from the Sir Elton John Collection, Tate Modern, London. November 2016. Author.

As if to echo the argument of the state of photography's afterlife, the exhibition *The Radical Eye: Modernist Photography from the Sir Elton John Collection,* held at Tate Modern from 10 November 2016 to 7 May 2017, exemplifies many of the temporal paradoxes of the modern in the space of the contemporary art museum and in photography. The exhibition unconsciously restaged an ancient pharaoh's tomb, although in this case it was the tomb of photography, excavated and opened to the public through the patronage of musician Sir Elton John, who since 1991 has amassed a collection of over 8000 photographic prints. The reasons for programming the exhibition were touched upon in a *Financial Times* review, where James Pickford[13] explained that since photography had acquired serious collectable status and prices in the art market, museums would find it hard to acquire notable works from the consecrated canon of photography. In Tate's case, whose commitment to seriously and strategically collecting photography only started at the end of photography, 2009 to be precise, they would have to rely upon the largesse of collectors to loan and eventually gift works to the museum. By showcasing Elton John's collection, Tate was underwriting a possible future bequest and establishing an agreement with a powerful collector. As the *FT* article went on to say, Tate was previously dependent on another collector; Michael Wilson, producer of James Bond movies, who had provided entire loan exhibitions as well as individual gifts.

The tomb of *The Radical Eye* comprised an entrance vestibule and six chambers, each of which contained rare and original photographic prints, framed in a wide assortment of ornamental frames, as they are hung in Elton John's own displays in one of his homes in Atlanta, Georgia. The liberal use of gold in the frames, in many cases hung together in groups, added to an overall sepulchral feel. This was an exhibition of the photographic crown jewels. The Tate curators had to work hard at modernist purification in setting out the exhibition and in the accompanying text, where it was necessary to reinforce key attributes of the modern. The exhibition notes emphasised that the pioneers of modernist photography were 'united by a fascination with the medium itself', and 'by stripping it back to its most basic components, artists celebrated photography, not as a tool for reproduction, but as a creative medium capable of producing new imagery'.[14] The exhibition is an alphabet of photography's now common formalist lexicon, which uses techniques of unconventional angles, extreme close-ups, foreshortening, contradictions of perspective and contrasts of light and form, which add up, as the curators say, to 'compositions hovering between abstraction and representation'.

In an *Observer* interview with Sean O'Hagan,[15] shortly after the opening of the exhibition, Elton John expressed the opinion that Britain had treated photography as a poor relation to art and his exhibition at Tate Modern was a vindication of photography as art. Elton John wanted people to react in 'awe and wonder' and to see 'the beauty of photography'. On the archived Tate web pages for the exhibition, Elton John links photography with his own musical formation in listening to traditional jazz, big bands, blues, rock and roll and country music. Photography and popular music are, says John, the art of modern times, or, as Bourdieu defined it in his sociological survey of culture taste in France during the 1960s, photography and jazz were middlebrow tastes (1996).

The puzzle of photography and modern times lies in its periodisation and its relationship to art. For Elton John the period covered by his own musical canon, by jazz, gospel and rhythm and blues traditions, coincides with the modernist works in his collection of photography, defined by and really exclusive to the first half of the twentieth century, but which is carried over to the present times. Elton John 'wants the public to look at Man Ray's photography and think, "oh my God, how beautiful is this and it was done so long ago, but it looks so modern, it looks so relevant"'. The modern aesthetic formalism of the first half of the twentieth century is transposed without a blink to the apparent modern of the present and in the present art can transcend its own period and manufacture in becoming timeless. The uneasy co-presence of the past and present, underwritten by modernism in the art museum, in *The Radical Eye* tomb of photography is finally undone by Elton John himself, in the same interview with Sean O'Hagan, when expressing his view of the digital. He says in the full quote, 'It all comes out of a computer now and I hate the deadness of it. You disappear up your own arse with digital technology. That's why I always record on analogue. I don't care about the charts, I just do it now. I can't get on the radio anymore and you know what, neither should I. I've had my time, I want to make records that I like in the way that I like, so it's analogue for me.' One more disavowal.

The Contemporary

So, the problem for the contemporary art museum is either that the modern has stopped being modern at some technologically periodised point and hence should be regarded as not speaking for the present, or that the contemporary continues as the logic of the present through disavowing technological change in order to maintain the modern. Either way it presents a conundrum of the inclusion of photography as heritage and the exclusion of the network. Returning to Claire Bishop, who introduced the idea of the art museum's fear of the digital, she says the contemporary is dysfunctional as a periodising term, but now functions as a discursive category. She quotes Peter Osborne, previously discussed in Chapter 4, who argues in his book *Anywhere Or Not At All* (2013) that the term contemporary is an 'operative fiction', designed to create a sense of unity to the present in the impossible attempt to combine disparate global temporalities. Osborne outlines the changing use of the contemporary in different contexts, from a common periodising term, designating that which is most recent, or up-to-date, to its more specific denomination in art history to distinguish the historical modern from the contemporary in art, to its postmodern inflection as a term of multiple temporalities. In this latter use, Osbourne emphasises contemporaneity as the present understood not simply as a coming together 'in' time, but 'of' times, in which time is increasingly characterised by 'a coming together of different but equally "present" temporalities, or "times", as temporal unity in disjunction, or a disjunctive unity of present times' (p. 17). It perhaps should also be noted that in *Being and Time*, Heidegger views temporalising not as the succession of time, but rather that the apprehension of time in the state of Being (*Dasein*) transcends the momentary event to a state where the past, present and future are co-present momentarily. This explains Heidegger's view that 'the future is not later than having been and having-been is not earlier than the present' (Heidegger, quoted in Wheeler 2011).

In less philosophical terms but those of 'contemporary' art discourse, the argument that Western historical time has been destabilised, with a consequent loss of futurity, means that art and its practitioners are stuck in a situation where the future's horizon has shrunk to that of the present, or as Bishop references Boris Groys's view, that contemporary art, and by extension its curation and exhibition, creates a non-historical excess of time. One of the cultural manifestations of such an excess of time is evident in the hyper-programming of all cultural institutions and its archival double on the Internet. For the discussion of photography at Tate it is also possible to consider that photography appeared in the 'contemporary' art museum at a moment when it was necessary to reinstate historical time, with the contemporary as a means of distinguishing the art of the present from the historical, which modernism could no longer do. In being purified and accepted into the canon, photography has assumed the role of illustrating historical time, which ceased to function in the contemporary because all photographs, even formalist ones, carry with them representational markers of time. Although able to reinstate or kick-start historical time, the problem of photography in the contemporary art museum remains stubbornly that photography is no longer continuous with a singular present. It is on this basis that time in photography exhibition can only be experienced as the loss of photography.

Pathos, Loss and Photography's Ruins

As if to prove the point that photography exhibition expresses the loss of photography, rather than the loss of time realised by photography, Tate Modern's *exhibition Conflict, Time and Photography* (26 November 2014–15 March 2015) amply demonstrates that it is photography, not human conflict, which is contemplated. It is possible to go further and say that what is curated is not photography's recording of human conflict, but the conflict within photography and the art museum. The exhibition was conceived and curated by Simon Baker, curator of photography and international art, with Shoair Mavlian, assistant curator, and Professor David Mellor of the University of Sussex. The curatorial framework for the exhibition was to select photographic images taken at different intervals during or after an event of war, from the immediate to its aftermath and commemoration, measuring not only the distance of atrocities from the present but the time lapsed between the event and the image. Notably the exhibition was timed to coincide with the 100th anniversary of the First World War in Europe.

Reading reviews of the exhibition from the time of its opening at Tate Modern in November 2014, it was this conflict within photography and not photography's realisations of conflict that was evident. Writing for *Frieze*,[16] Sara Knelman commented, 'This wasn't, ultimately, a show about specific conflicts, or even conflict in general. It was about photography's relation to these events: about distance, trace and memory, and photography as a mode of remembering, reflecting and abstracting'. It is a short extension from Knelman's observations to saying that the not fully disclosed subject of the exhibition was the mode of photography, photography's past. The exhibition's imagery could be read just as much as a history of the medium, and the formal, pictorial and technical preoccupations of photographers over that period, as the intention of its curators to show photography's relationship to historical time. Roger Fenton's 1855 image of a cannonball-strewn track in a desolate landscape, given the literary title 'Valley of the Shadow of Death', marks the early moment of photography's participation in war, at a time when photographic technology was too cumbersome to reach battlefield events and camera exposures were too slow to capture fast-moving actions. At the other end of the history of the medium, Adam Broomberg and Oliver Chanarin's 'The Day Nobody Died' (2008) consists of exposed sections of a large roll of photographic paper to the sun, fixed and displayed as a unique C-type print of 762 mm x 6000 mm, showing abstract streaks of black, orange, blue and white. The artists made this exposure while embedded with the British Army in Helmand Province as war artists and chose to make a nonrepresentational work, or a work which called into question photography's collusion in recording cruelty of war in representational terms. But, as Sean O'Hagan, reviewing this work at the time for *The Observer*,[17] asked, what, by refusing to document the war by means of conventional photojournalism, did they achieve, 'what did they offer in its place but a postmodern performance where abstraction – the exposed sheets of film – and mundanity – the photographs of a box of paper – converge, creating a series of photographic gestures in which images are emptied of meaning to make a rather laboured point?' As with other works of Broomberg and Chanarin, the medium of photography is challenged, through continued use of the medium of photography, in order to point out the limits, if not redundancy, of photography. What replaces photography in such works is the work of art. Broomberg and Chanarin's 'The Day Nobody Died' is the most obvious and conceptual work in the exhibition's trope of abstract formalism, which challenges photojournalism. Laura Cumming in her *Observer* review[18] notes that 'this show stints on human beings themselves in favour of emptiness, ruin and absence,' arguing that the consequences of war and its weaponry are more convincing when shown through injury to bodies. She argues that Kenneth Jarecke's image of a burned Iraqi soldier in his tank on the road to Basra would not be included in the exhibition because the photographer is a reporter, not an artist, and points to what she sees as an arbitrary distinction between photojournalists and artists, with the exception of the inclusion in the exhibition of Don McCullin's 1968 image of a shell-shocked American soldier. Here is the conflict within photography, between art and photography and in its undisclosed terms the conflict within the mode of photography. The lack of disclosure leads eventually to the conclusion that photography exhibitions exhibit not only Barthes's 'that which has been,' which means that every photographic image is an image of death, but now that the exhibited photographic image is also the 'that which has been' of photography. As if to underline this, Sara Knelman concludes her review by observing the danger of an impotent retrospection which aestheticises the horrific and an abstraction which hides atrocity, as if to say photography is not up to the job, because in the art museum it can only contemplate itself. As Walter Benjamin saw the situation,

> The destructiveness of war furnishes proof that society has not been mature enough to incorporate technology as its organ, that technology has not been sufficiently developed to cope with the elemental forces of society. The horrible features of imperialistic warfare are attributable to the discrepancy between the tremendous means of production and their inadequate utilization in the process of production – in other words, to unemployment and the lack of markets. Imperialistic war is a rebellion of technology which collects, in the form of 'human material,' the claims to which society has denied its natural material. (Benjamin 2008 [1939])

Jonathan Beller extends Benjamin's argument in arguing that photography has been part of the technological means which produces war, which, in Benjamin's terms, would mean that it is important to locate war photography and more generally photojournalism as an excess of the social potential of the means of production in the form of a rebellion of technology. This makes it impossible for photography to comment upon this excess, unless its practitioners are also able to comment upon the mode of photography's reproduction, of which for the most part it has shown itself incapable in its numerous reproduceable forms.

Don McCullin Exhibition at Tate Britain (5 February–6 May 2019)

Something of Benjamin's historical tragedy of capitalist technology manifest in war is evident in Don McCullin's reflections on his life as a photographer. Tate has produced a number of short videos for McCullin's exhibition, both for Tate Britain and for the modified version at Tate Liverpool. At the beginning of one of the Bloomberg Tate Shots films, a short black-and-white hand-drawn animation, McCullin, now 85, is heard saying, 'The majority of the last fifty years of my life have been wasted photographing wars'. In another of the films available online to Tate Members of the Liverpool exhibition, McCullin comments that his work hasn't changed the tragedies of war because 'as soon as one war finishes another starts'. Since the Cold War in Europe, the permanent arms economy and the military–industrial complex have ensured the continuation of proxy wars fought for control over the extraction of the resources of the earth. Currently there are approximately 40 active conflicts. Millions upon millions of people have died violent deaths as a direct outcome of war since the end of the Second World War, and according to the United Nations 90 per cent of people killed in these conflicts are civilians. This is the context which has left McCullin reflecting on time lost in his life, the events of his own life which drew him into photographing conflict and suffering and the failure of photojournalism in the face of the structural perpetuation of war. In a Tate Members video of the Liverpool exhibition, McCullin says he hates being called a war photographer. He also says, counter to the description of him by Tate, as well as by his dealers Hauser and Wirth, that he is not an artist but a photographer, a title he is happy with.

The exhibition was curated for Tate by Simon Baker, director of the Maison Européenne de la Photographie in Paris, with Shoair Mavlian, director of Photoworks, assisted by Aïcha Mehrez, assistant curator of contemporary British art at Tate Britain. The exhibition consists of 250 black-and-white photographs McCullin has taken over the last 60 years. It includes images from conflicts in Germany, Cyprus, Biafra, Congo,

Vietnam, Cambodia, Northern Ireland, Bangladesh, Lebanon, Iraq and Syria as well as images made of poverty and homelessness in London's East End and the people and landscapes of the post-industrial North of England as well as more recent landscape images in Somerset. The Tate Britain gallery walls were painted a sombre grey for the exhibition. With the exception of a digital slideshow of colour images that appeared in *The Observer* and *The Sunday Times* magazines, McCullin's photography is the world in sombre black-and-white. The exhibition prints were made by McCullin, dodged and burned in his own darkroom, accentuating the high contrast and the darkness of the medium's tonal range. In McCullin's own words, 'What I hoped I had captured in my pictures was an enduring image that would imprint itself on the world's memory'. The black-and-whiteness of McCullin's fine prints is also the enduring memory of photographic-ness on display in the exhibition. The overwhelming narrative of the exhibition, woven around Don McCullin's life and work as a masterful, self-taught outsider and almost accidental photojournalist, is one of loss for a world that might have been, a more promising and hopeful world, and a testament to the flawed world that happened. The modernist exhibition form aesthetically marshals and reshapes McCullin's body of work. In curating it for the art museum, images originally made for the print industry, itself radically changed by technology over the past three decades, are recontextualised within Tate's modernist project of committing itself to photography, stripped of their located and contractual context, apart from McCullin's voice. As expressed, Tate's commitment is to modernist and contemporary photography, which is a temporal paradox, since the modern is now historical and the contemporary can no longer claim a unified purchase upon the present. The result, as has been argued, is photography now as a memory in the same manner as its images. In this respect the Don McCullin exhibition ('the image imprinted on the world's memory') is not only a mourning for the world as it might have been, but a requiem for the medium of photography.

6

Photography and Heritage: A Case Study of the Victoria and Albert Museum

Celebrating the slightest object from the past, invoking the duties of memory, remobilising religious traditions, hypermodernity is not structured by an absolute present it is structured by a 'paradoxical present,' a present that ceaselessly exhumes and 'rediscovers the past.'

Gilles Lipovetsky. 2005. *Hypermodern Times*. Cambridge: Polity, p. 57.

Over the past two decades, heritage has acquired a new and expanded dimension, and while this has been seen locally as a consequence of the financialisation of cultural organisations, there is a more complex logic, related to the erosion of Western linear time and its replacement by the multiple temporalities and mode of exchange of global technocapitalism. It is in this context that photography now has the status of heritage, a medium of the past, but paradoxically is still taking part in the representation of 'the past as heritage' by means of digital reproduction. It might be said that this is photography as 'living heritage,' a fitting term for photography's zombie condition, which can be discerned here as an uneasy coexistence of the past and present, hovering between life and death, still sentient in its 'undead heritage' condition. It is this very paradoxical condition which leads Tate Modern to collect and exhibit photography as contemporary and modern and leads the Victoria and Albert Museum (V&A) to continue to extend its historical photographic holdings as 'the art of photography.' This might simply be thought of as two complementary versions of heritage, and indeed the modern and contemporary at Tate is now heritage, but that is not how it is understood. At Tate, photography is distinguished from heritage by the singular time of the present and by aesthetic modernism's aesthetic purification. What follows is a further case study of the appearance of the afterlife of photography in the museum as a means of examining the troubled taxonomy of collecting photography and what it reveals about photography's relationship to historical representation.

Photography at the V&A is included within a history of the designed object and collected as heritage, by means of what are considered the finest examples of its technical and aesthetic manufacture. Paradoxically, once more, the distinction between heritage and contemporary collection is observed between institutions, while being dissolved within them. The lack of distinction between contemporary and heritage collection has resulted in a convergence of galleries and museums of all kinds commissioning and acquiring new hybrid art works, purified by collection and exhibition, drawn from analogue photographic, film and video sources, but whose default is digital reproduction. It might be argued that technical convergence has simply created a parallel condition of cultural convergence, in the form of new artistic media. Are museums and galleries not simply recognising and operating within this new state of affairs? Yes, indeed the V&A is continuing to collect and exhibit works which it sees as expressions of contemporary sensibilities, as reflected in the media artists and photographers are using. However, the problem examined here is with how the relationship between the past and present is presented in the V&A's photographic displays. Curating photography according to linear time leads to the acquisition and display of what is considered significant contemporary photographic work, which both reconfirms the historical collection and projects a continuous expanding future. But in the hypermodern commodified world, there is much less confidence in the future and its horizon has shrunk to that of the present. In such a situation the cultural authority of curating is also challenged because the distinction between a recognised past and a confident future has broken down. If this is the case then cultural value has been changed, raising the question of how the 'everything' of culture is to be curated if everything has the status of heritage.

Institutional collecting is a selective process, based upon an availability of means, together with knowledge and expertise and carried out according to historical tradition, in which cultural value is both expressed and defined. Collections are made on the basis of the best of a cultural form and/or to be representative of a type or class of object. Objects of collection are understood to embody ideas, reveal new understandings and generally play back and reflect an age or period. But this settled process has already been undone and overtaken by the Internet and by the museum's entry into digital data and corporate instrumentality. Museums are caught in the paradoxical position of resisting and embracing this new set of conditions and the big questions are how the Internet is changing the cultural value of the objects it digitises and how and what museums should collect in and of computational culture.

The chapter draws upon the launch of part one of the V&A's new Photography Centre in 2018, as a way of seeing how curatorial authority is pursued in the paradoxical present. With the digitisation of photographic collections, the longstanding distinction between collection and archive is put under great strain, if not collapsing altogether. Many of the questions about photographic collections and archives raised in this chapter stem from this fact alone. Some consideration needs to be given to understanding the relationship between an analogue photograph and its digital 'copy' and then to understanding the materiality of the data file and cultural value of the digital image. The digitisation of collections began experimentally in the late 1990s and has since become a standard procedure for museums and over the last decade an established expectation of audiences in accessing collections online. Over the period in which museums have been digitising objects of collection, professional questions have arisen over the selection, cataloguing, provenance and copyright, over what has and hasn't and what can and can't be digitised. Alongside these curatorial problems, digitisation entails technical problems of rapid developments in image scanning and capture equipment, alongside greater computer power, leading to constantly shifting standards of digital files, which in turn raises questions of the preservation of digital artefacts. The digitisation of heritage is an unwieldy and impossible project of the encyclopaedic imagination, which nevertheless is propelled by data as the new currency of knowledge.[1] The implication for digitised heritage is that with the projected scale of data, no human could encompass nor comprehend such a deluge and therefore heritage data is ultimately designed for machines to read and see. For photography such problems impact upon how the historical collection of analogue photographs is made accessible, as well as how the museum will collect digital-born images in the future.

Making the contents of museum collections accessible online was recognised by Google from 2009, which saw the potential in offering its Streetview and gigapixel camera technology, software and technical support to museums in return for exclusive agreements to host and market museum collections. The Google Art Project was launched in 2011 with 17 participating museums, a figure which has grown to over 2000. Google understood the public relations value of a project which hosted the world's heritage as a gift to the world, but how it conceived and carried out that project bears more scrutiny here. What is included in this chapter is by no means a thoroughgoing analysis of Google Arts & Culture, but it is used as a specific example of how photographic images, as distinct from other art and craft objects, are rendered on the interface screen, in which the differences between the archival object, its digitised form and appearance on screen can be discussed. The starting point of all of this, as well as the basis for the underlying analysis, is to consider first what is meant by heritage in the present context and how it structures the experience of photographic images.

Heritage as Commodity

As Lipovetsky observes in *Hypermodern Times* (2005), the heritage being celebrated today, and moreover consumed everywhere, is not seeking to permanently inscribe memory on to the very sites and objects of the past in order to maintain a continuous connection with tradition. What is being considered here is the cultural modernisation of heritage, an ultra-modernity, or hypermodernity, by dint of the formidable expansion of what is considered worthy to belong to the memory of the past. The signs of this obsession with commemoration are manifest in a number of forms: heritage visitor centres using visualisation techniques to conjure the distant past; the regeneration of post-industrial towns and cities; the expansion of world eco-tourism; and the expansion and popularisation of museums. The mass democratisation of heritage is expressed not only in visitor centres and museums but in broadcast television's appetite for historical costume drama and documentaries whose subjects are every conceivable object of the past, amplified by spin-off publishing and domestic lifestyle design. This, then, is heritage in which the past is consumed in ephemeral moments of celebration, in diversifying leisure and filling up time. The economic impact of the cultural modernisation of heritage has taken the form of privatisation in which the priorities of income generation replace the spirit of public service. What Lipovetsky is calling here cultural modernisation, which he characterises as ultra-modern, is of the same order as Ulrich Beck's (1994) analysis of reflexive modernity, defined as the capacity of capitalism to rescue itself from its own unbridled excesses through elevating reflexivity to an organisational level in post-traditional society. In political terms, as the historical nation state objectively weakens against global capital markets, heritage functions to reinvest continuity with the past and shore up diminishing monocultural identities in imaginary and mythical narratives. The presentism of hypermodernity reduces collective affinities with historical objects at the same time as individuals are able to consume and possess more objects. It is a case of the excesses of an all-consuming present in the face of anxieties about the future, and a proliferating inflation of memory, expressed by the boundless broadening of the frontiers of heritage.

The Disciplinary Museum

In both Tony Bennett's *The Birth of the Museum* (2013) and Eilean Hooper-Greenhill's *Museums and the Shaping of Knowledge* (1992), the development of the museum is explained in Foucauldian terms of the disciplinary society, arising from the politically reforming and largely secular sciences of the Enlightenment and the social enlargement of the public sphere. In such accounts the museum functioned to educate and civilise a new and growing public, by virtue of its invested cultural authority, based upon scholarly disciplinary knowledge. The view of the museum as a disciplinary social institution, formed primarily in the nineteenth century, led to the critique of the museum in the late twentieth century as out of step with a social democracy in which culture was newly defined as a shared property of everyday life, rather than a quality to be legislated by an educated and disciplining 'elite'. As discussed in the previous chapter, by the end of the twentieth century Tate had successfully adapted to encompass both continued minority scholarly interest in its collection with a wider popular appeal to the public enjoyment of art based upon experience, rather than disciplinary knowledge. The V&A certainly emerged in the nineteenth century from the reforming interests of its first director Henry Cole, combining public enthusiasm for the massive expansion of machine-manufactured objects in the first half of the nineteenth century with an expressly educational role in improving standards of taste. By virtue of its founding the V&A has been closer to everyday life and popular forms of consumption than fine art institutions or antiquarian museums. The V&A was founded in a period of a still-expanding Empire, reflected in popular interest in world geographies, where photography functioned to return the Empire to its centres of operation by means of pictorial representation. The post-colonial critique[2] applied to the museum was a response to the questionable means by which historic objects from colonial exploit had been collected, as well as the unexamined forms in which such objects were displayed. Such concerns remain active and ongoing and expressed as the need to both repatriate objects as well as the need to decolonise museums and develop new narratives of colonial exploitation and subjugation. As a recent example, the Black Lives Matter campaign against continued institutional sanctioning of racism has reminded museums of the need for continued work on their historical narratives of display. However, the critique of the museum, from a post-critical perspective, plays out in the hypermodern heritage museum according to a different logic, one which is based upon consumption and the paradoxical present in which competing aims coexist within the organisation of the museum and its practices.

Tristram Hunt, the current director of the V&A and a former shadow secretary of state for education in a Labour shadow cabinet, is clearly aware of the sensitives of the current political moment for the museum. Hunt sees the postwar tenets of liberal democracy crumbling and so it befalls the museum, along with other civic and media organisations, 'to preserve the ecology of civic life' against 'the growing militancy of identity politics' and 'divided politics and emboldened autocracies' (2018). Museums have a responsibility as, 'trusted arenas of public space, civic leadership and intellectual credibility'. Hunt treads a liberal tightrope in believing that 'museums still have a public role to play in telling complicated stories of cultural hybridity and cosmopolitanism' and with a 'rigour, transparency and curatorial knowledge base that can only help to foster an educated citizenry' (2018). Hunt is not too far from the traditional mission of the museum here, even as the V&A develops organisational practices of heritage based on leisure and entertainment in which marketing becomes the curatorial centre. How this current moment in the museum, particularly at the V&A, bears upon the question of the afterlife of photography is the point to pursue here.

What can be said at the outset is that the history of the technical development and widespread application of photography from the 1840s is closely associated with the history of the commissioning and acquisition of photographs at the Victoria and Albert museum, which opened to the public in 1852. Both photography and the museum were shaped by and participated in nineteenth-century colonialism, with its disciplinary urge to dominate, extract, classify and collect, and the parallel politically reforming domestic disciplinary movements across Europe, necessary to discipline a class into wage labour, which eventuated in modern educated and enfranchised social democracies. The world was surveyed by a Eurocentric analogue photographic image technology and in turn the world surveyed as a photographic image was returned to European museums, emphasising Britain's superiority in trade and manufacture.

Photography: An Uncollectable Medium

The *Great Exhibition of the Works of Industry of All Nations* of 1851 in London brought together an international collection of over 100,000 objects of the arts of industrial manufacture and raw materials, including photography, which was also used to document the exhibition building and its contents. The *Great Exhibition* marked a developmental moment for photography in the museum, functioning in the greater organisation of museum data and in the production of wider publicity in print. The exhibition was a commercial success, visited by over 6 million people, the equivalent of a third of the UK population at the time, and made a profit of £186,000, which was used to purchase land in South Kensington for the building of the Victoria and Albert Museum. The first ever dedicated photography exhibition in any museum in the world was held in the V&A's upper refreshment room in 1858, organised by the Photographic Society of London and consisting of over 1000 photographs, including

250 contributions from its French counterpart, the Société française de photographie. Henry Cole, the first director of the V&A, who promoted the art of photography as well as being an amateur photographer, recorded in his diary that Queen Victoria and Prince Albert attended the photography exhibition private view. Cole began collecting photographs at the South Kensington Museum, now the V&A, in the year it was established in 1854.

With the transfer of the Royal Photographic Society Collection from the National Museum of Science and Media in Bradford in 2017, the V&A photography collection now holds 800,000 photographs dating from 1839 to the present day. The first phase of the new Photography Centre at the V&A presents a display from its enlarged collection, which contains a continuous and unifying narrative of photography. The narrative is stitched together from the different periods, interests and motives for collecting 'the art of photography', or collecting photography as an 'independent art', as Mark Haworth-Booth, curator of photography at the V&A between 1970–2004, outlines in *Photography: An Independent Art* (1997). The many uses of photographic reproduction, the photographies met at many points across this volume, were historically reflected in the distribution of photographs across different departments of the V&A. There was no one unified photography collection, but dispersed groups of photographs lodged in different museum archives. As Haworth-Booth charts, there were a number of attempts in the postwar period to establish a stronger profile for photography through lobbying for a national centre or a national collection of photography at the V&A, which were ignored by directors at the time. It was not until 1977, under the directorship of Roy Strong, that photography was brought together in the newly created Department of Prints, Drawing, Photographs and Paintings, which began a ten-year programme of contemporary photography acquisition as well as backfilling its historic collection. Interestingly enough the desire for a national British collection of photography expressed itself again in 1987 with the formation of a Committee of National Photographic Collections, comprising representation from 11 museum collections of national importance: Birmingham Library, the British Film Institute, the Imperial War Museum, the National Buildings Record, the National Library of Wales, the National Media Museum, the National Portrait Gallery, the Public Record Office, the Royal Photographic Society, the Scottish Photography Archive and the Victoria and Albert Museum. What it evident is that photography has been held by these institutions, in archival collections, defined by their content, context and purpose in which photography was regarded as the technical medium. Only comparatively recently has this order been reversed in seeing the medium as the unifying feature of dispersed and highly differentiated bodies of images. Why do these distinctions matter in understanding the heritage value of photography? Isn't the heterogeneity of photography an essential part of what photography is? The answer being pursued here is that in the light of digital reproduction the relationship between archive and collection of the 'photo-object' is crucial in unpicking the linear master narrative of photography, reflected in the opening display of the new centre, *Collecting Photography: From Daguerreotype to Digital*, which, as the director Tristram Hunt said, will 'seamlessly span the entire history of photography.'[3]

The V&A's Photography Centre

In the same manner as the V&A's first ever photography exhibition of 1852, the private view of the new Photography Centre in 2018 was attended by Kate Middleton, the Duchess of Cambridge, who unveiled a plaque to mark its opening, thus preserving an enduring relationship between the V&A, the monarchy and photography. The opening display, *Collecting Photography: From Daguerreotype to Digital*, was presented as exploring photography as a way of 'collecting the world.' The strategic decision to give greater prominence to photography at the V&A in a phased development is underlined by the transfer of the Royal Photographic Collection from the National Museum of Science and Media in Bradford in 2017. As Tristram Hunt said, the transfer 'provided the catalyst for this dramatic reimagining of photography at the V&A.' The seamless span of photography's history was achieved in the opening display by a number of curatorial means, but primarily by the commission of two new 'photographic' works by German photographer Thomas Ruff and American artist Penelope Umbrico.

Ruff selected from the V&A collection the waxed paper negatives of Captain Linnaeus Tripe, an official photographer for the East India Company, made over 160 years ago of temples, palaces and monuments in India and Burma. Ruff, who as the V&A commentary suggests, is interested in the science of photography, was drawn to the way in which the discolouration and damage to the paper marks the passage of time, as well as being interested in Tripe's early retouching process of painting on the reverse of the negatives to add different effects. Ruff made a series of large C-type prints, entitled 'Tripe | Ruff' (2017). Of this work Ruff has said, 'In the age of digital photographs, I find it really interesting to revisit these images', and that, 'Tripe | Ruff' 'is about the history, the different processes, techniques and technology of photography, and how rich the photographic world really is'.[4]

Penelope Umbrico produced a site-specific video work titled, '171 Clouds from the V&A Online Collection, 1630–1885', displayed on the centre's newly installed grid of nine screens known as the 'Light Wall'. As with Ruff, Umbrico uses images from the collection, selecting details of painted clouds from the online collection of paintings. The process of making the work involved cropping hi-resolution scans of paintings, importing them into video authoring software's pre-set 'colorizing' filters and applying slow dissolves between each image. The resulting digital video is 56 minutes long and shows a moving cloudscape. Umbrico has said of the work,

> Like the cloud, photography on the web is inherently fragmented, un-assignable, un-locatable, indeterminate, de-contextualized, and constantly changing. But when printed, photographs are a reflective medium like the painted clouds in the photographs of them that I am using. But I am interested, here, in the fact that filtered sunlight is replaced by the filtered electronic light of the screen. So perhaps this is not an *analogy* of the current state of photography, but rather, it might *be* the current state.[5]

What these two works achieve for the V&A is precisely a seamless transition between technologies and mediums across time, which happily maintains the continuity of the V&A's historical narrative of collecting and displaying the art and science of photography. However, in doing this, the technological seam between the analogue and digital, what is taken here as a radical disjuncture between the photographic and computation, remains unseen. But the seamlessness is intended in the artists' ploy, especially in the 'Tripe | Ruff' work, in which the original historical analogue negatives of Tripe were photographed on a light-box at the London museum and sent to Ruff as digital files, which he then digitally

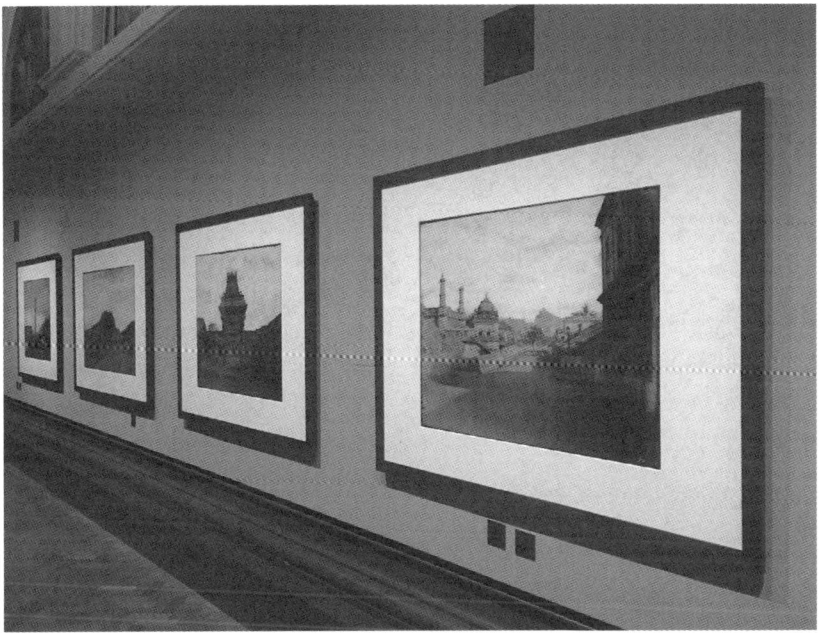

Fig. 6.1

Thomas Ruff, *Tripe* | *Ruff*, the Photography Centre, Victoria and Albert Museum, London. January 2019. Author.

manipulated by overlaying the negative and positive image. The final digital images were exported and printed as analogue C-type prints and traditionally mounted, framed and hung in Room 101. A neat technical and historical trick had been performed between two technologies and two distant periods in the history of photography – the digital returned to the analogue, the analogue treated digitally. In the video screen work of Umbrico, presented as the inaugural commission for the centre's 'Light Wall', erected as the means for displaying future born-digital 'photography' works, the analogue/digital trick is performed in reverse. The work the V&A has acquired is a 9.27 GB .mp4 file (3840 x 2160 px), which can be displayed through a combination of software programmes which can 'play' the digital file and display it across a grid of nine screens. The technical journey in this work is from an analogue painted surface to the museum's digital files, sections of which are digitally copied into a digital

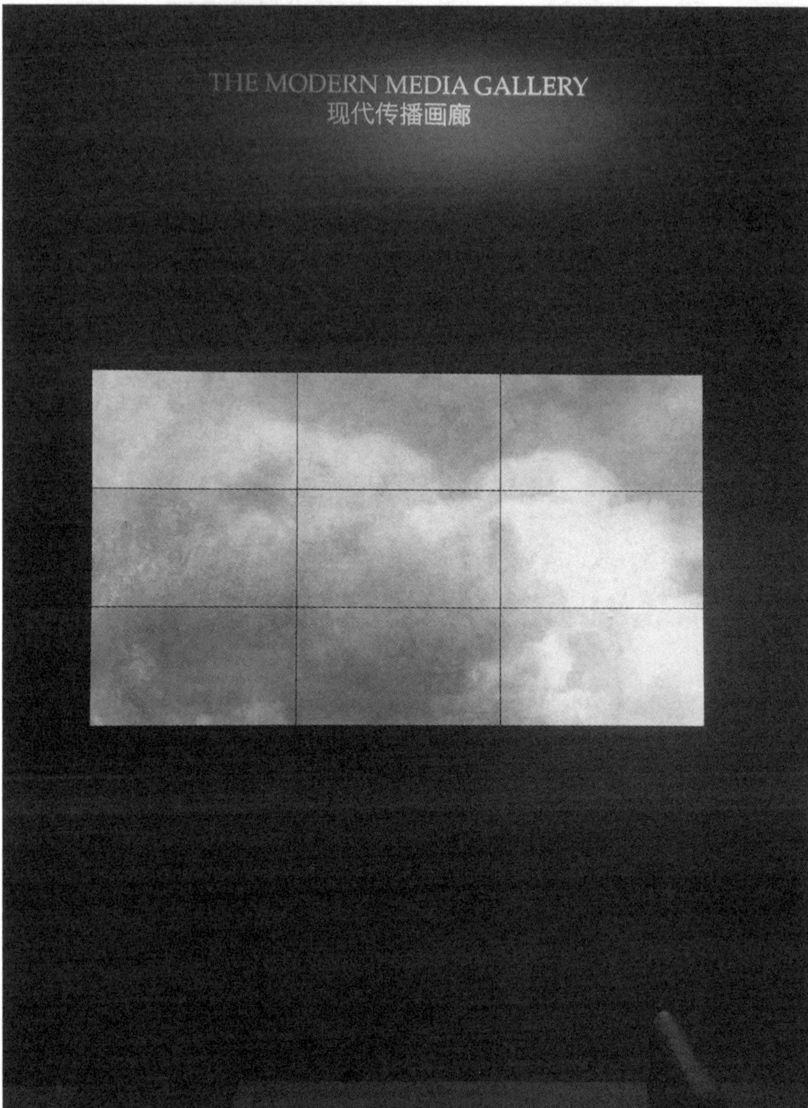

Fig. 6.2

Penelope Umbrico, *171 Clouds from the V&A Online Collection, 1630–1885*, Modern Media Gallery, Victoria and Albert Museum, London. January 2019. Author.

video authoring software programme, producing transitions between the images, which are then edited into sequences and digitally screened. Whereas Ruff returns the digital manipulations made on screen to chemical prints, Umbrico takes the reflective light of an analogue painting as a reference for the production of a digital manipulation of projective light.

The digital mode of production, the software employed in both works, remains secondary to the resolution of the images in print and on screen. These are works whose subject is not the new technological medium of the image, but rather the artistic image of photography produced by a new medium, whose subject is the analogue in painting and early photography. This is clearly the point of the commissions, but it reveals an absence of curatorial interest in the art and science of the medium being used, which is the art and science of the computational image. The digital and computational is the default of image production, but rendered invisible by the desire to keep photography going at all costs. The logic of serious collections is that they have to continue to grow and in order to do this they have to encompass new developments in a specialised medium and cultural form. New acquisitions are the lifeblood of collection, while dead collections ultimately stagnate. What would the V&A's Department of Prints, Drawings and Paintings add to its photography collection if not photographs?

Archives

In a report for the Comité des Sages of the European Commission on the 'The Cost of Digitising Europe's Cultural Heritage' prepared by Nick Poole in 2010,[6] European Museums alone house 350 million photographs suitable for digitisation, and among the approximately 30 million individual photographs held in national libraries about 4 per cent had been digitised at the time of the report in 2015. By 2020, the British Museum had reorganised its online collection database, making over 1.9 million digital images of its collection available online for free use under a Creative Commons License.

Archives, rather than photography collections, contain by far the majority of the world's photography heritage, discounting here the widespread possession of family photographs, and with digitisation the heritage value of archives is being increasingly realised. In the UK alone the 11 organisations that comprise the Committee of National Photographic Collections, hold millions of photographs, accumulated in different contexts, for different classificatory reasons. The British National Archives, whose photographic holding runs into millions of individual items, amassed its collection over the working lives of central government departments over the historical time-span of the medium itself, gathered through the business of government in commissions, the office of copyright, collected by British diplomats overseas, inherited or donated. The subject matter of the photographic images details the governance of the state infrastructure in transport, housing, industry, crime and war. The National Portrait Gallery has more than 250,000 original photographic images, of which at least 130,000 are original negatives, dating from the 1840s to the present day. The British Library holds a number of specific archival collections, many in illustrated book form, including photography and notebooks in the Fox Talbot collection, the India Office photography collection of images of India and Asia from the 1840s onwards. The Canadian Copyright collection consists of some 4000 photographs registered between 1895 and 1924. The cartographic collection, also in the British Library, consists of a series of nineteenth-century topographical views. Twentieth-century photography is represented in the British Library's Fay Godwin collection and the Kodak Historical collection. Historic England holds over 9 million photographs, the Science Museum Group holds photographs from medicine, science and industry as well as by amateur photographers and the National Trust has a large collection of photography across its historic houses, including the Fox Talbot Museum at Lacock Abbey. The same situation is broadly mirrored in most European countries, accounting for the 300 million photographs suitable for digitisation, and this situation is no doubt replicated in many other countries, including the USA, where in addition to national offices, universities and museums have acquired archival collections from other parts of the world. But what the impossible project of digitisation of photography illustrates is not photography's ultimate unity as a medium, but precisely its disunity and dispersal, reinforcing yet again the understanding that there were many photographies and that the idea of photography is generated only in relationship to its discursive context. Such an understanding rests upon John Tagg's (1988) analysis of the second technical revolution of photography at the end of the nineteenth century, in which he argued that photography has no identity outside of the institutions in which it is practised. For the discussion of the terms in which photography is collected as heritage, the point remains that photographic archives play a part in mobilising heritage narratives as well as having documented social behaviours of heritage tourism. But the photography of the archive is not the photography of collection, and the boundary between them is policed precisely because it can never be settled.

Reproducibility

One very obvious point to make along the way is that the reproducibility of the photographic image and photography as a medium of reproduction make it a universally available object of collection, and in theory not a commodity investment. The photograph can be reproduced to infinity and like any mass-produced object is affordable and widely possessed. Photography became institutionally collectable from its inception as the property of a selective group surrounding its early practitioners, which gave it a marketable value. But even here there is a question of what the property of a photographic image is; the image is in one sense free-floating because of its reproduceability and therefore it is the image's inscription on a surface, in a negative or a print made from a negative, which constitutes the property – in fact it is the analogon which is given a collectable and marketable provenance. As photography entered the art market more fully from the 1980s, ownership of a photographic art work also took the art market form of limited editions or contracts to limited ownership of specified reproduction of images.

Interest in the physical photographic object of historical collections and attention to its material form, as a negative, print, slide, as stored in boxes, folders, albums, together with accompanying textual annotations, seems, initially at least, to be the opposite of Walter Benjamin's observation of the radicalising effects of photographic reproduction. In current academic interest in the archival photo-object, photographs are taken as unique objects, ironically reintroducing the idea of an aura attached to unique works of art, whereas, as Benjamin pointed out, photography obliterated aura through reproduction. From the perspective of gaining a synoptic view of a medium of the past, there is, initially at least, something of an Alice in Wonderland effect in attempting to read the social and political process of reproduction from the object form of the image, yet it can and is being done. In *PhotoObjects: On the Materiality of Photographs and Photo Archives in the Humanities and Sciences* (Bärnighausen et al. 2019), Costanza Caraffa reminds us that the physical presence of the photograph bears traces of its use. In this respect Caraffa says that 'photographs lead a double existence as both pictures of objects and material objects in their own right,' and as objects they are accorded agency in playing a role within the knowledge practices in which they were deployed. Elizabeth Edwards, in 'Thoughts on the "Non-Collections" of the Archival Ecosystem' in the same volume, argues that through the materiality of photographic practices it is possible to discern an invisible eco-system, a 'non-collection,' beyond the bounds of the archive, operating as a hidden logic, which disturbs, 'the hierarchies of value and categories that have created collections and performed photographs as certain kinds of things' (2019, p. 67). As with other forms of analysis developed from actor network theory and media archaeology, objects can and do give up their secrets, through the material process of their practices, which play an active part in meaning making structures and systems. But, from the perspective of forgetting photography, which insists that the default of the techno-social agency of the image has moved somewhere other than photography, such archival studies approaches can only be taken as reinvestments in reproducing the category of photography. And, as with every example looked at so far, art and media historians, artists and photography curators, view the digital and computational as a continuation of the medium. Studies of the materiality of the analogue photograph lead 'seamlessly' to studies of the 'born digital photograph,' an oxymoron from the perspective of this book, and add to the ever-expanding list of academic research topics as well as collections. Studies focused upon the materiality of the photo-object are productive in leading out into analysis of unwritten or implicit codes of institutional archival practices and should support the case for considering the present situation beyond photography, but ironically they lead to recasting photography in the image of its past. The 'digital-born photograph' is not another kind of photograph, but an image temporarily realised in a data stream, which belongs to a nonrepresentational system. The double existence of the photograph as image and object, a picture of an object as well as an object in its own right, relies upon an analogue system of representation which doesn't carry across to the computational. It is possible when considering the analogue to make the separation between representation and reproduction, the 'picture of the object' standing for the representational function and the 'photo-object' for the function of reproduction. With computation the image is simultaneously and inseparably reproduced in the nonrepresentational algorithmic code, and at the same time represented as a visual data object, simulated as a photograph which is instantiated in a dynamic and relational transmission, of human to human, machine to machine and between humans and machines. Such a definition clearly needs further detailing, taken up in the final chapters, but the point here is to arrive at an understanding that the human comprehension of the computational image relies on the representational code of the photograph, whilst its reproductive function doesn't, and that attempts to continue to study the computational image as if it were a photograph miss the point.

Meanwhile, back in the V&A's new Photography Centre, the institutional curators, charged with making and displaying a collection of historic and contemporary photographic images according to the narrative of the art of photography, are working within a cultural discourse which unifies the object of photography, allowing it to stretch back and forth across the timespan of the medium and the museum. Such curatorial positions don't recognise the argument that the medium has been replaced. If they did, it would open up both a new historical vista upon the limits of what photography was and what it did and would begin to articulate the new ways of computational and seeing. The curators of photography at the V&A might argue that the inaugural exhibition of the centre does show what photography is and does, and that the argument made here is, at best, splitting hairs, or altogether wrong. The counter-argument is that the distinction between the post-photography of the analogue/digital photography on the one hand and the beyond-photography of the computational networked image on the other hand matter to both the way the history of photography is revised, as well as how contemporary images are understood. There is a wider socio-political need to challenge the accepted epistemology of Western representation and its worldview that in turn is related to understanding what is happening in the digitisation of the photograph and its circulation in online collections and how the museum presents such understandings to the public.

Photography and the Google Cultural Institute

Google Arts & Culture is an online platform which makes items of collection from leading museums and archives available in digitised form. Google states that its Arts & Culture project is a non-profit initiative, which originated from its policy of encouraging employees to develop projects of interest within work time. Its mission statement is 'to preserve and bring the world's art and culture online so it is accessible to anyone, anywhere',[7] a slogan reminiscent of Coca Cola's marketing strategy in the aftermath of the Second World War, when the American Army committed itself to providing its troops with access to Coca Cola. At the launch of the Google Cultural Institute in Paris in 2013, its director Ahmet Sood was quoted as saying that 'In order to understand culture you first need to be able to experience it. Some people are lucky enough to visit a museum and see a work of art in person but not everyone can travel the world and see different cultures and culture. Quite simply, the internet can help to export local culture to a global audience, allowing anyone that has an internet connection to explore a heritage site or virtually jump inside a gallery as if they were there.'[8]

The platform was launched in 2009 by the Google Cultural Institute with the participation of 17 international art organisations, including Tate, MoMA and the Uffizi. By 2012 the platform contained 34,000 art works from 151 museums and art organisations from 40 countries. By 2020 Google Art & Culture was available in 18 different languages, with over 2000 institutional partnerships. The model used by Google was to offer its image capture technologies to digitise museum objects for free, in return for partnership agreements allowing Google to host selected works. Google used its technology, consisting of its panoramic Streetview camera, lasers used to capture the distances to walls, motion sensors to track the position of the portable cart containing the camera, a hard drive for data and a laptop to operate the system, to produce 3D-navigable renderings of exhibition galleries. The data gathered was subsequently synced with Google Maps. In addition Google used its gigapixel camera to scan 2D works in large-file-size sections which were then amalgamated using software to produce images of very high resolution. The platform is powered by Google technologies, using a Java-based Google app engine and a web application using Google APIs. The interface is adapted for mobile or desktop screens and can be searched using 13 search terms as well as a selective navigation of 'highlights', showcasing the different technologies used. Google Art Project returns 205 collections, 2679 stories and 327,532 items, 6 museum views and 396 user galleries, highlighting different aspects of photographic collections. Photography can also be searched as 'the medium of photography', which returns 36,800 items. Photography on the Google Arts & Culture platform has much the same problem as museums do in classifying photography across different uses and contexts, appearing as a topic in its own right, as contemporary documentation of cultural heritage sites and events and as selective archival historical events. All of the 327,532 items are displayed onscreen as transparent images in an interface design functioning as a media player, something similar to the Netflix interface, to make content intelligible, searchable and clickable, tempting the user to stay on and return to the site via constantly changing topics, tasters, places to visit, stories to be told and wonders of the world to experience. The real problem with photography on the Google Arts & Culture site, as it is with any digitised image, is what the appearance of the photograph on screen signifies. This needs some consideration.

Digitisation of the Photograph

The V&A has a partnership with Google Arts & Culture, displaying on screen over 5000 items from its collections of textiles, costume and jewellery, from the seventeenth century to the present. The online collection also contains photographic images that have been selected to illustrate styles and types of clothing worn in the period, what Barthes called the 'studium' of the photograph, or its averaging effect. Among the photographic items reproduced are *cartes de visites*, magazine images, individual prints, as well as digitised pages from John Thomson's book *Street Life in London*, published in 1864, made by a photomechanical reproduction process using pigmented gelatin of a purple-brown colour, which gave a high-quality look and permanence to the finished images. The digitisations of these pages are facsimiles of the original pages on which the historical photograph is framed on the page. In other digitised images the photograph appears with a thin black border or is shown without any framing. The photographs are digitised in colour, which reproduces the type of manufacture, whether in colour, sepia or black and white.

It is hard to hold in mind that the 5000 or so V&A images on the Google Arts & Culture platform are all photographs of a kind. The difficulty arises because the on-screen image has a transparency, inviting the viewer to look through the screen. The transparency of the on-screen image is designed for immediacy, to let the viewer experience the object shown by means of the cultural conventions of photography. The mediation of this transparency, its digitisation in fact, is not visible and not referenced, whereas the hyper-mediacy of the interface is present and conventionalised in now-familiar modes of navigating content (Bolter and Grusin 1999). There is no convention to register the computer simulation of a photograph, which displays a transparent image. Computing has developed technically on the cultural assumption that embedding 'rich' media content is a natural aim of transmission and reception, that music, film, television, text and photographs are piped through a digital converter. The digital channel is considered a technical tool and its development is aimed at greater immediacy, to get as close as possible to the experience of the object in the museum and of being in the museum, without being encumbered or made aware of the technical media which gets you there. In 3D authoring software, which uses photographic rendered images, the surface of the screen and the screen's image are fused in perception, functioning as a transparent window through which a perspectival world of objects are recognised. This effect of the screen's embedded image-producing transparency has an additional meaning for the digitisation of photographs, because analogue photographs operated on the same basis of the transparency of the image. In this equation, digitisation foregrounds the objectness of the analogue photograph, its bounded surface qualities, flaws and deterioration, whilst concealing the objectness of the simulation of photography through which the transparency of the screen is produced.

The Digital Remembrance of Photography

Whilst Google's image algorithms are the engine of nonrepresentational information, the Google Cultural Institute and Google Art Project naively simulate the photograph, thus ensuring the continued zombie state of the photographic image. One of the consequences for the photograph, also based upon the logic of transparency, is a double transparency that disappears the object, that is, the photograph. The result is that the photograph is rendered as a digital image with the exact equivalence to all other screen images, i.e., whatever it is that is photographic is lost. The tension between the dispersal of the practices of photography into general computing and the increasing aggregation of analogue photography in archives, collection, exhibition and their organisational systems represents a new moment of an older problem of the identity of photography. The advent of large- and small-scale high-definition flat screens in fixed and mobile combinations, together with increased digital file sizes, greater storage capacity, transmission speeds and networked databases, now brings the screen and image, in what have been their dominant forms – television and photography – into a new embrace. Taken together, the proliferation of images, screens and the network represents a new moment of accelerated convergence which is not only evident in commercial products and new forms of media distribution, but is becoming visible in the educational and public arts sector, through independent digital media organisations and projects exploring the curation of images in the network.

The Google Cultural Institute shows us the obvious flaws in taking culture as a given and the interface as a transparent channel. Perhaps it was too easy a target, but it is salient in pointing up the problem of one of the predominant ways in which the relationship between art and technology, or photography and technology for that matter, has been seen. To regard the technology as simply the channel or the tools is to ignore that network interfaces are deeply value-coded as well as mathematically coded. What the computational image shows us is that 'the real' is more occluded than ever. The afterlife of the photograph stands in the place of private intimate life on the one hand and a negotiated public life on the other. Software and profit have sabotaged both.

If such a state of affairs as I've described were to be granted, then it brings about the need to reconsider what the digitisation of objects of collection actually achieves for understanding heritage. The passage of time along which the more singular idea of digital culture has travelled has itself been overwhelmed by the very practices it originally called forth, such that now there is a multiplicity and reduplication of image-knowledge hybrids, circulating in networks, which confound attempts to maintain linear historical accounts and singular objects. Such a situation in which the digital itself has a history as well as constituting the mode of production makes the position of continuing to regard photography as a contemporary medium even more untenable and produces not only a confusion of the representation of photography's history, but also adds to the confusion over the current status of the image in network culture. The network image is what lies beyond photography and frames Part III of this work.

Part III

7

The Image after Photography

Certainly progress in photography is not going to stop. Faster material and lenses, more automatic cameras, finer-grained sensitive compounds to allow an extension of the minicamera idea, are all imminent. Let us project this trend ahead to a logical, if not inevitable, outcome.

Consider a future device for individual use, which is a sort of mechanized private file and library. It needs a name, and, to coin one at random, 'memex' will do. A memex is a device in which an individual stores all his books, records, and communications, and which is mechanized so that it may be consulted with exceeding speed and flexibility. It is an enlarged intimate supplement to his memory.

Vannevar Bush. 1945. 'As We May Think', *The Atlantic*, July.

The Image after Post-Photography

This chapter attempts to move beyond photography in considering the current state of the visual image, but inevitably stops short on the central argument of the book, because representation is still based upon the photographic lexicon, paradoxically simulated by the nonrepresentational system of computing. Nevertheless, the chapter sets out a series of 'transitional steps' in the formation of the image after photography, from analogue inscription to big data, from image apparatuses to social performances and from the discourse of photography to the discourse of computing. This is a journey from reflected light striking a chemical surface to the metadata of an image file and from the image as a unified object to the image as a relational field.

While it is the case that the image still has countless individual material instantiations, it now operates at volume and speed, which is more defining of its character. The increased scale, rapidity and transience of the production, circulation and reception of images suggests that the image is better defined in the social sense as a transmedial flow (Henning 2018). Mobile Wi-Fi and smart phones afford the image spatial mobility, creating interactions between bodies and subjects, and give the image a dynamic, affective, performative and transactional character. These new characteristics of the image foreground its relationality and participation as a socio-technical assemblage. The technicity of the image is also at issue here and following John May's (2019) genealogy of the current image as mathematical and electrical, existing as signals in real time, this chapter reinforces his technical argument that photographs and the (digital) image have virtually nothing in common. The case for developing a socio-technical relational definition and field of study of the image stacks up on both sides of the scale, in which the technical, following Bernard Stiegler, is deeply connected to consciousness and thought and, following Bruno Latour, where the social is conceived in relationship to the agencies of technical apparatuses. It is of course possible from the current techno-historical vantage point to remap nineteenth- and twentieth-century photography in the same relational terms, and indeed this would be a better way of seeing the history of photography, as an image technology inseparable from the organisation of the means of production and domination (Beller 2018). As the post-photography discourse has shown, photography also continues to be mapped forward, as evidenced in some of the technological qualifiers of the image examined below, raising again the now perennial question of what purpose is served by retaining the language of photography. This question surely breaks on distinctions which need to be made between an historic medium, imaging technologies and the modes of their operation. Can it be said that photography continues to stand as an engine of the mode of visual reproduction, or is it computation and the mode of informational reproduction? Is the remediation of photography an equivalent to the representational image, or is the computational image already a different image, a hybrid of graphics, animation and cinema – and, if so, aren't a new set of terms needed to account for this image?

What then is the image freed from photography? As a passing reminder, nothing is lost of photography in its forgetting; rather, as was suggested earlier, there is much to be gained by photography remembered by its rewriting. However, the problem for any account of the new condition of the image remains the residue of photography, its afterlife, which cleaves to the image through custom and practice. Participation in the networked

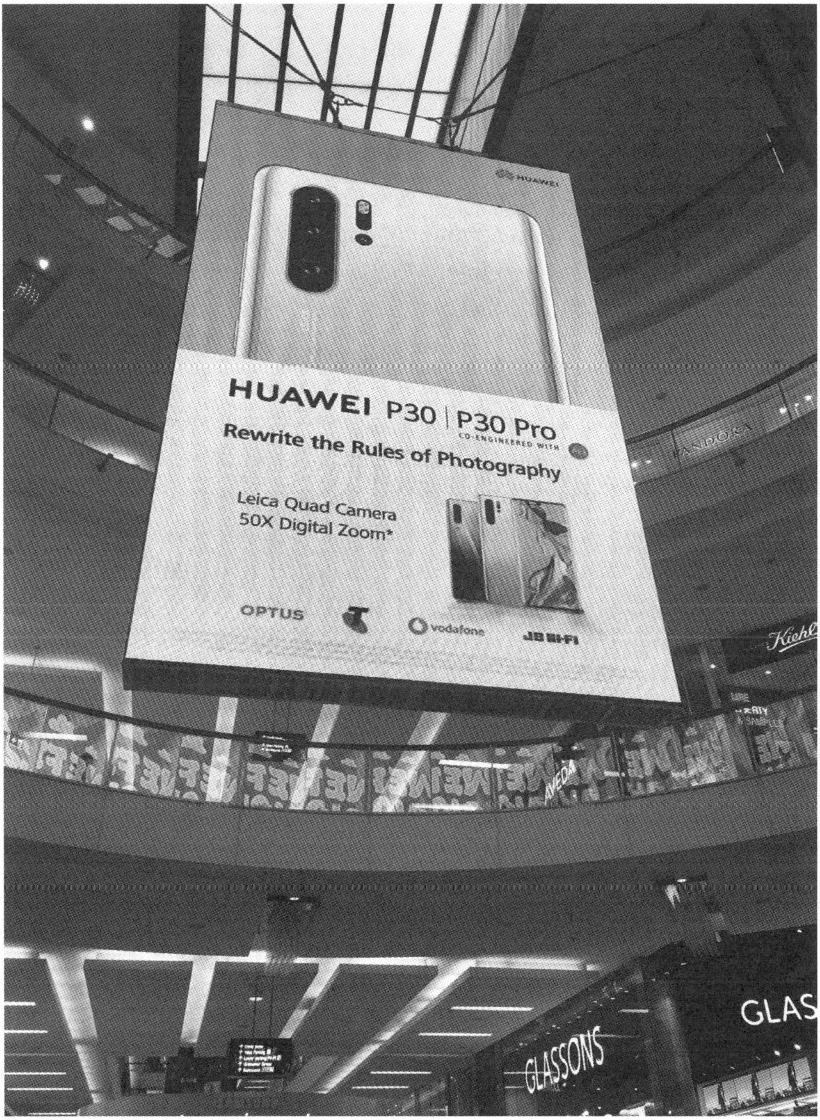

Fig. 7.1

Video screen, Westfield Shopping Centre, Bondi Junction, Sydney. 2019. Author.

image is increasingly experienced as a performative act in everyday life, but it is still articulated in the older language of photography, even as a new set of terms for image culture is emerging in the language of phoneography. The terms cross with older terms of photography, bound to the leisure commodities of camera and smart phone equipment, in which the computer in your pocket is marketed as 'rewriting the rules of photography.'

Photography remains a commodity market and a commercial profession, with an obvious vested interest in keeping the image photographic. As has been said in numerous contexts, more people make more images than ever before, stored in remote and secretive server farms each the size of several football pitches, circulated online and populating the interfaces of the Web, but whether the number of commercial photographers has risen or fallen with the new condition of the image is a mute sociological point. Throughout the history of photography, the occupation of photographer has been itinerant, changing with technical apparatuses. Now, contemporary technical economies require a substantial labour force of eyeballs employed to scan, process, sort, click, delete and report on images, but it is unlikely that any of these operatives consider they are involved in photography. The image, as a data package in informational logistical relay, is inimical to the contemporary global mode of production in which monitoring and surveillance are central, but this is not the photographic image. In most practical ways the severance of the image from the photographic has already taken effect, and the cultural debate about the future of photography is but another lingering aspect of its residual form. But it would be a mistake, as has been discussed, to think of the technical change from analogue to digital image in terms of the continuity of the image in culture, just as much as it is a mistake to think of the death of photography being wrought by digital technology. The digitally generated, computationally processed and networked screen image is neither a photograph nor a digital image, but, paradoxically, is still received as one or the other. In severing the link between what belongs to the image and what belonged to photography this paradox of a representational system of signs and symbols shimmering on the surface of a nonrepresentational screen becomes clearer.

The screen image is a product of sensors, software and code, which bears the imprint of its artifice, even as software aspires to photographic verisimilitude, which is more and other than its technological production. The visual appearance of the screen image might then be considered

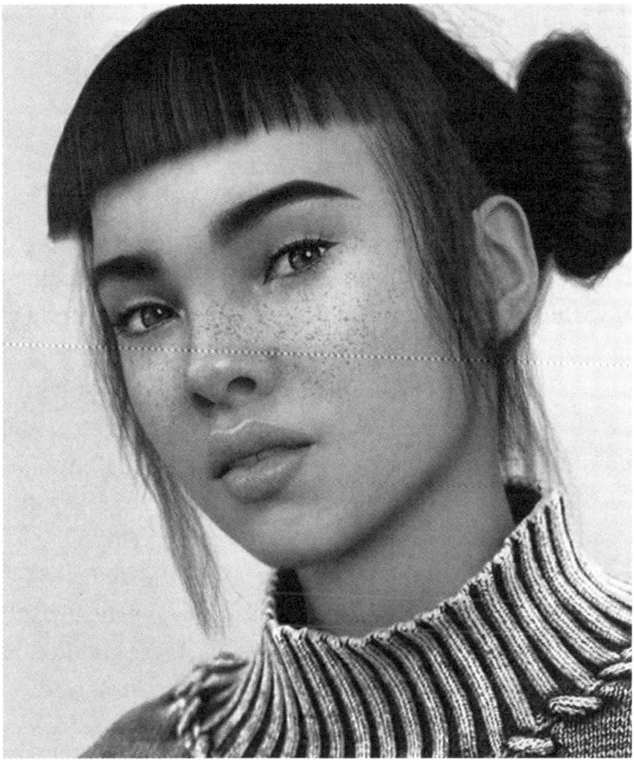

Fig. 7.2

Digital model Lil Miquela Sousa, screenshot, thecut.com, *The New Yorker*.

as a hybrid SFX video animation in its technical compositing and as a contingent screen graphic in its reception. The look of the new image is painterly in depth, colour and contrast; it is programmable and mutable; it can move and it can be still. This is a visual image that flows across the screen of digitally remediated cultural forms of television, advertising, print and cinema and its polysemy is read through symbolic transcultural narratives. The aesthetic of the new image might well be considered a fashion image and an image of fashion.[1] A clear example of the new visual features of the image are to be found in virtual influencers, notably Lil Miquela, who first appeared on Instagram in April 2016. Miquela Sousa, or Lil Miquela, is a 19-year-old Brazilian-American model, musical artist and influencer with over a million Instagram followers, who is computer-generated.

The image of Lil Miquela in 'selfies' has all the traits of the composite and transmedial. Her appearance draws most obviously upon media characters from Disney animation, the Star Wars films and fashion magazines, pulled together in a convincing CGI avatar with an extended cross-platform media life and narrative. Prada worked with Lil Miquela to promote its Fall 2018 collection via animated GIFs on Instagram Stories and fashion magazines, including *King Kong* and *Paper*, have photographed her, working with a press agent to organise the photo shoots. Lil Miquela is not alone in either the fashion industry or computer animation studios; this is a modelling world in which the filters of Photoshop were once used to make real models 'fake', and in which sophisticated animation software can now make 'fake' people look real. Shudu is a virtual Black supermodel created by former photographer Cameron-James Wilson using 3D animation tools DAz3D and Clo3D rather than photography, which he regards as archaic.[2] The generation of virtual influencers is a development of 'real' media celebrities, or personalities, whose lives fill the demand for perpetual content on social media platforms. Social media is hybrid of the real and the virtual, a prosumer space for the mixing of fact and fiction in the constitution of communities of interest surrounding lifestyle consumption. Social media is a complex virtual communication channel in which fashion products and advertising engage with users in sophisticated campaigns of brand awareness and allegiance. Fashion is a cultural site of a politics of identity representation, in which Lil Miquela's has been deliberately composed as a representative type, a forward-looking new generation of mixed-heritage Americans, a market the fashion industry cannot ignore.[3] Lil Miquela is a convincing animated virtual model made by a financially savvy media company based in Los Angeles, using a generation of powerful computational tools which can create and synthesise photo and video and insert them into any digital media in far less time and at far lower cost. This is clearly reflected in the online investment company TecChrunch's website reporting in 2019 that Lil Miquela's creators, Brud, had closed a $125-million investment round led by Spark Capital.[4]

The new visual image is both more real than real and more obviously less real than any real. The new image is hybrid and hyper-real and while the representational code of visual equivalences still relies upon the cultural convention of photography, the new image is essentially a graphic animation. Software replicates, enhances and increasingly predicts the form of imaging based upon photographic assumptions. However, it has been pointed out that analogue photography was also a prescribed and mediating system and that the differences between analogue and digital are less than their similarities; indeed, in post-photographic discourse the light fixing of the analogue photographic image has been revisited as proto-digital, as an argument for the continuity of the photographic image (Zylinksa 2021). But such comparisons become highly attenuated when moving beyond the visual aesthetic of the new image, to its dispersal and transmutability across networks. Here the new image is coded and transmitted by non-visual and nonrepresentational means. In informational terms the new image is data, but in representational terms it has to be regarded as transmedial and this is a crucial aspect in understanding its meaning and value, a topic discussed in relationship to education, mediacy and literacy in the final chapter. But this is to get ahead, and the starting point for the new image is to consider its condition after photography.

A Final Note on the Photographic Image

In considering what is revealed of the image after photography, the question of the vestigial link between photography and computing is worthy of a final note. Most ontological investigations into the essence of photography converge on the phenomena of light capture. Light has remained the defining quality and principle of the photographic image; Fox Talbot's *The Pencil of Nature* (1844), Roland Barthes's perfect analogon in *The Photographic Message* (1961, in *A Barthes Reader*, 1993) and Jean Baudrillard's *Photography, or the Writing of Light* (2000) bear out the continuity of light capture as the irreducible core of photography. In the context of all contemporary photo media, light is considered its unifying and in another example the capacities of light-based technologies and their techniques are traced back to the photographic (Cubitt et al. 2014).

The photographic image is the fixing of a pattern of reflected light, passing through an aperture for a duration of time, to fall upon a two-dimensional plane. The dark chamber of the camera obscura allows humans to witness the nature-technical phenomena prior to its 'graphic' treatment, as does the prism of the camera lucida. However, as Baudrillard observed,

> The light of photography remains proper to the image. Photographic light is not 'realistic' or 'natural.' It is not artificial either. Rather, this light is the very imagination of the image, its own thought. It does not emanate from one single source, but from two different, dual ones: the object and the gaze. 'The image stands at the junction of a light which comes from the object and another which comes from the gaze' (Plato). (Baudrillard 2000)

Light belongs to both photography and the human gaze. In the image considered as a relational field, the physics of light (photons and waves) is but one part of the equation, the other being embodiment. In subtracting photographic light from the image, which was the imagination of the image, the configuration and contingency of the image is opened out once more to the light of the gaze and imagination before and after photography.

However, the continued centrality of the photon/waveform image is still pressed in photographic theory, notably recently in Joanna Zylinska's *Nonhuman Photography* (2017). The inscription of light is also central to Susan Schuppli's concept of material witness, an exploration of the evidential role of matter as both registering external events and exposing the practices and procedures that enable matter lurking in grains, pixels, molecules, particles and in the blurry outlines of x-rays or the fuzz of radiation to bear witness (2020). Zylinska proposes that, in the context of the timescale of possible human extinction, photography can be taken as part of the broader natural-cultural history of the planet, through which it is possible to trace parallels between photographs and fossils, and 'read photography as a light-induced process of fossilization occurring across different media' (2017, p. 104). Zylinska argues that by virtue of its non-human technical form, that is, by seeing its non-human operations, photography is not a technology opposed to nature or culture, but rather reveals the energy and light of nature and life itself. In this way Zylinska positions photography as part of geological time and the deeper time of the material inscription of light, which is both digital and analogue. Zylinska uses Batchen's (2000) account of proto-photography in *Each Wild Idea*, in which he says, 'If we read digitality as the intelacing of an ON/OFF pattern that is translated, or rather transcoded, into different material media – light sensitive papers, silicon chips and so forth – we can see that photography, even at its very inception, reveals itself to be always already digital.' From the perspective of arguing beyond photography, Batchen and Zylinska stretch the concept of digital code to mean any pattern produced by the alternating absence and presence of light, which Batchen relates to an electrical transmission recalibrated as the zeros and ones of mathematical code. In Batchen's view of photography as 'always already digital' and Zylinska's photography as fossilisation there is a desire to reject technological obsolescence as a narrow and determinist view of the history of photography. They do this by expressing the continuing desire for photography and by arguing for first and natural causes, in effect arguing that in the beginning nature was digital. Isn't arguing that the digital image is an outcome of the purpose and design of analogue photography guilty of a teleological view of technological development?

The light photography brought under its control, the light bent by the aperture and lens, snapped by the shutter and inscribed by light-sensitive chemicals, is a formal restriction of humanly visible light and less than the light of the restless bifocal mobile bodily gaze. It is an historical and cultural form of light, as the imagination of the image. The chemical or electronic fixing of patterns of light is a predetermined grading formula, a scientific idea embodied in a technical apparatus and system (Flusser 2011). The camera obscura light reflected image and its permanent chemical registration remain wonderous discoveries in the human/nature continuum; however, they continue to span separate moments in time, history and culture. Photons and waves are necessary to the materiality of the photographic, but the image is a product of the scientific and cultural imagination. The captured light which produces the photograph is not the dangerous excessive light of the electromagnetic spectrum and the universe, and the algorithms which govern the registration of the light pattern of the digital image are equally not relaying some 'natural' light. In the computational image, light has already been calibrated before its entry into the camera. As John May (2019) argues:

> Photography and photo-detection are held apart from one another as technical categories by this fundamental and un-bridgeable epistemic abyss between heliography and bolometry, between photography as written light and imaging as detected energy, between Nicéphore Niépce's View from the Window at Le Gras (ca. 1826) and John Logie Baird's telescan of Oliver Hutchinson (1926). (p. 13)

Mapping the Field of the New Conditions of the Image

In attempting to arrive at an understanding of the image after photography it is necessary to review the various ways in which the image has been technically accounted for over the last three decades in the context of the instrumentalisation of vision and its applications. Over this marked accelerated period scholars have attempted to keep up with conditions and above all to define the image around a cluster of technologies, which connect the transmission of vision with data collection and storage. Such technological developments have been propelled, as was the Internet, by a military–industrial complex, and the permanent arms economy (Kidron 1989). Research departments and laboratories were fed by defence spending and contracts and focused upon technologies which would enhance command, control, intelligence and operations. A related consequence of the military–industrial complex is its symbiotic relationship to the production of commercial products and services justified as spin-offs. Communication technologies form a bridge in research and development across military–industrial and commodity production and as such have clearly impacted upon the changing landscape of media production and consumption. In the development of vision technologies the connection between military, state, civil and commercial applications has recently become evident in applications of facial recognition software, data security and remote guidance systems. In scholarly and artistic attempts to understand such technologies terms such as digital, algorithmic, computational, operational, non-human, network and soft have been applied to the formation and functions of images. Within such developments critical scholarship and investigative journalism have identified a unifying drive towards automation through artificial intelligence, with all the questions this raises for the future of society and the planet (Bridle 2018).

Trevor Paglen, in an *e-flux* article titled 'Operational Images' (2014), refers to Czechoslovakian-born filmmaker Harun Farocki as one of the first artists to draw public attention to the application of the computer image in weapon technology. He referenced the guidance system of cruise missiles as early as the 1980s and realised that images were no longer being used to represent an object but were instead part of an operation. In his video installation trilogy 'Eye/Machine III' (2003), Farocki named these images operational: images that track, navigate, activate, oversee, control, detect and identify. In effect operational images are instrumental in performing tasks. Paglen went on to make the point that operational images have become invisible and are not made for humans, but by machines for machines to read. The algorithmic and operational are two powerful descriptors of the computational image as a non-human agent functioning within data systems applied to drones, autonomous cars, medical imaging and automated industrial assembles. Nathan Saucier (2017) in his MSc thesis reinforces Paglen's view that machines are performing inscrutable assessments in interpreting images and providing conclusions for which the reasons remain opaque.

The autonomous dimension of computational imaging has been defined by Joanna Zylinska as non-human photography (2017). Zylinska creates a new category of the non-human in photography, by identifying three categories of photographs. First are the frequently encountered, uncanny-looking photographic images which are not *of* the human, her example here being of depopulated expansive landscapes. Second are photographs that are not *by* the human, such as images produced by traffic cameras or Google Streetview. Third are photographs that are not *for* the human, including QR codes and algorithmic modes of machine communication that rely upon photographic technology. Her purpose in defining non-human photography is to move beyond humanist representational photography, aligned with dualist and binary epistemologies, in order to reposition contemporary photographic practice with the politics of the post-human. Unlike artists such as Paglen and Farocki, Zylinska sees collaboration between the human and machinic as creative and fertile. For Zylinska the autonomous and computational image are not antithetical to human photography, but open up a new relationship between photography and technology in which photography is rediscovered as a force of nature from which a new ethics of vision and practice unfolds.

The Formation of the Image after Photography

Taking a step back from the political urgencies of questionable technological applications of vision-entailed technologies, it is now necessary to consider the components and sequential development of the image after photography. Mapping the formation of the image and its technical nexus will help inform the final two chapters, which return to the politics of the image and its imbrication in everyday life. What follows is something of a genealogy of the image after photography outlining a sequence of critical understandings in moving from the photographic image to its current computational default.

Much has been written about the digital image since its first commercial appearance in the 1990s and its general principles have become widely known, with many digital photography websites giving detailed explanations.[5] For the purpose of considering the image after photography, the significance of the digital image lies in two aspects. Of principal importance to all subsequent developments is the original technical development of the digital image, converting continuous analogue signals into discrete electrical signals, registered as a mathematical value. A digital image is produced by a rectangular array of pixels known as a bitmap in which each pixel represents a colour and tonal value at a single point in the image, which is stored as a number. It is the binary code which makes the image compatible with computation and indirectly led to new image capture devices. Digital camera sales have declined by over 80 per cent since 2010, mostly in the market for compact cameras, due to the rise of smart phone and mobile 'photography' apps. Smart phones use the principle of digital conversion by using a light sensor and an image signal processor. The sensor, a complementary metal-oxide semiconductor (CMOS), captures light photons as they fall upon the surface of the sensor, made up of tiny photosites, which are converted into electrical signals of varying strength depending on how many photons were captured by the photosite. At this point in the capture process the potential of the image, or in older photographic terms its latent capacity, has been registered. The image signal processor (ISP) converts the first set of data into a digital image by converting the electrical signals into a colour and tonal value for each pixel, recorded as a number.

Historically, the significance of the digital image lay in its contrast with the photographic image and early discussion centred upon whether the digital image could achieve the same image resolution as film. As the technical standard of digital imaging improved, debate shifted, as is well known, to questions of the veracity of the digital image given the manipulation made possible by code. But today the significance of digital imaging, and the second point of interest here, is no longer in the comparison with the analogue, but in how electronic imaging has enmeshed with computing and what that enables. Technical developments in digital image capture have advanced considerably in both spectrometry and computing to the point at which light capture and processing produces the variable data image. Research in lensless image capture technology uses permeable film imprinted with specific fringe patterns, instead of a lens, which does not form an image but projects a shadow pattern made when light passes through the permeable layer. Lensless cameras combine an optical technology of projecting shadows and a signal processing technology of converting the shadows into captured images. The software used by the processor has been trained to create an image by comparing patterns produced by sensors of cameras with a lens to patterns captured by the sensor without a lens and pairing this with the information coming from the film mask.[6] Similar is the development of high-dynamic-range (HDR) imaging, which captures multiple images of the same scene using different exposure values which are synthesised by software into a single image that represents the full range of tonal values across the light patterns captured. The Light Field Camera, developed by Lytro over a six-year period and marketed in 2012, developed another image capture technology which combined the intensity and direction of light rays crossing a given plane, also known as the 'light field' at that plane, producing an image which could be refocused, as well as changing the point of view when viewed on a 2D screen. While Lytro ceased trading in 2018, Google started R&D in light field technology, reputedly employing Lytro engineers. The point is, as histories of media technologies amply demonstrate, technical development produces outcomes, not all of which eventuate in successful market applications, but which indicate a more general direction of interest and intent and in some cases are taken up in new combinations for different purposes.

As advances in light sensors and RAM speed continue, smart phone and smart camera technology use increasingly sophisticated models, trained by machine learning to decode and assemble images. Trained software can recognise and recall every element of an image, and can also predict elements in the making of images. Such developments have over the past decade marked a paradigm shift in image capture, which can be summarised as a change from the optical to the computational, further severing the link with photography. The photographic image is now only replicated as software, in which case it can be said that the afterlife of photography is memorialised by a computer programme, while on the other hand the digital image becomes ever more permeable as data.

A Quickening of the Pace towards the Computational Image

In their seminal paper 'The Digital Image in Photographic Culture: Algorithmic Photography and the Crisis of Representation' (2013a), Daniel Rubinstein and Katrina Sluis used the term algorithmic photography to argue that the combination of software and image was not merely a different processual image, but a paradigm shift with major implications for the ontology of the image. They went on to point out that in computational photography the image's resemblance to an object is a matter of the algorithmic processes that operate on the light pattern collected by image sensors and that the same data could be outputted as a text file, a sound or a string of numbers. The algorithm comes between the raw sense data and an output, in what Rubinstein and Sluis, quoting Eivind Røssaak, call 'an algorithmically enabled work process'[7] that humans can read culturally as a photographic image. In describing the nonrepresentational character of the algorithmic image they suggest that it contains two competing logics, a rational visual representational logic whereby the image corresponds to an external object and a 'recursive viral logic of intensity, multiplicity and incompleteness in which the image only refers to itself' (2013, p. 35). For Rubinstein and Sluis, the algorithmic image is 'Like a two-faced Janus' which 'points in two directions at once: one side faces the objects, people and situations as they appear in the "real" world, and is occupied with the representation of events by flattening their four dimensional space onto the two dimensional plane of the photograph. The other side points towards photography's own conditions of manufacturing, which is to say towards the repetition and serial reproduction of the photographic image.' What was retained in this analysis of competing logics, and other understandings of the computational image at the same time,[8] was the idea of photography itself, which still leads to something of an impasse in considering how the computational image is to be met in cultural ways of seeing. From the perspective of forgetting photography, the logic of Rubinstein and Sluis's important account of computational photography eight years ago remains a riddle precisely because of the paradox of photography and representation. They were pitting the algorithmic against representation, what subsequently developed into the idea of the nonrepresentational and later non-human image. But if the algorithmic image is no longer a photograph, why retain it, and if the algorithm serially reproduces the photographic image, then what is the algorithmic image? The logic of the argument of the algorithmic image was directed against the prevalence in photographic and media studies of structuralist and semiotic methods of image analysis. There was more than good cause in pointing out that images were no longer singular, static or linear, but rather processual and therefore not amenable to an essentially art historically derived form of analysis. Rubinstein and Sluis's analysis fruitfully raised more questions than it could answer at the time. A question pertinent to this discussion is how the processual image is to be understood in terms of its reception. How in effect can the computational image be seen, if not as a photograph?

The computational image is the opposite of the direct, integrated inscription of the optical image. The computational image is indirectly formed by using discrete measurements from different sources and points in time. The computational image is formed through algorithmic processing, which integrates an array of information. Computational imaging has applications in microscopy, tomography, ultrasound, seismology, cosmology, passport control, traffic regulation as well as 'photography'. As algorithms of computational imaging become 'trained' to teach computers to see and apply their knowledge independently of their human operators, computational imaging will be employed in governance and everyday life. Such algorithms will be used in law enforcement, criminal detection and maintaining social order, as well as more diffuse consumption in which computers will be even more intimately involved.

Cultural analysis of the computational image at some point inevitably reaches the boundary of computer science, and for the culturalist, stepping over into the world of mathematical code and electrical signalling requires a transdisciplinary knowledge that so far few hybrid culturalists or computer scientists have achieved. Transdisciplinarity will be necessary in order to gain a complete knowledge of the computational image in society and one way it can be achieved is by collaborative research in which the technical knowledges of culture and computing can come together to formulate new questions and goals. The field of software studies is one such attempt to understand that computing is cultural and the cultural is computational and is an important contribution in expanding and reframing humanities perspectives (Fuller 2008).

What lies behind the continual expansion of technical knowledge in image capture and computational imaging and what drives such developments? A naive answer might simply be human inquisitiveness, the ever-expanding frontier of knowledge about the physical universe and the apparatuses needed in order to see, record, measure and communicate. There are many other more practical answers to such questions, such as the proliferation of consumer devices, in this case functions of cameras and smart phones, which are faster, smaller, lighter, cheaper and automatic. This is certainly what lies behind lensless technology and image sensors. Technical innovation is also driven by more veiled scientific, military and industrial interests in wanting to gather information at higher resolution, at greater speeds and over greater distances. Scientists, engineers, military strategists, product developers, designers, investors and civil manufacturers all have specific and sectional interests and rationales for working upon image capturing devices. Taken together all of these reasons add up to something greater and more abstract. Sectional interests are shaped by the general mode of production, based upon capital accumulation, which directs not only research in image capture, but the entire technical military and civil apparatus, what Jonathan Beller earlier called the substrates of computational capitalism. With the digital image and computing, the historically distinct media apparatus of photography dissolves into a greater system of technologies and what emerges is the computational image, whose goal is automation.

A specific set of issues for the image after photography surround how the computational image and vision algorithms come to be written, how are they trained, who their authors are and what values are built into the algorithms. Automation through computing is popularly depicted as the autonomous operation of machines, taking over previous human functions, the robot of science-fiction being the popular imaginary. The autonomy of the computer is sought in the goal of artificial intelligence, which it is assumed will arrive at the point at which machines teach the next generation of machines what they know, without human intervention. But such a view of the future of AI, always just around the corner, ignores the collaboration between humans and technology. This is not to underestimate the increasing power of computers to process data at a scale impossible for the human mind, but it is a necessary reminder that data is dependent on its source and in the sphere of the visual image, many of those sources remain human. Understanding how computer vision programmes have been developed is a case in point. Nicolas Malevé has been undertaking collaborative research into machine vision with The Photographers' Gallery for the last four years, examining the training data for computer vision algorithms through a focus upon the method and conditions in which photographs are annotated at speed and at scale. His paper 'On the Data Set's Ruins' (2020) distils and translates a much larger body of research represented in his unpublished PhD thesis, titled 'Algorithms of Vision: Human and Machine Learning in Computational Visual Culture' (2021).[9] In his research Malevé examines how the photographic image is mobilised as an instrument to transform the visual into data, where data of different origins can be compared and classified. Malevé carefully unpicks the development of computer vision, showing how visual input is unproblematically aligned with the representational code of photography. Further, he describes the shift in computer vision research, from early explicit modelling of pattern recognition to current techniques based upon the use of large-scale visual databases, in which machine learning algorithms detect pixel pattern regularities by comparing examples. The important point for Malevé is the large scale of data required for machine vision and examines the case of ImageNet, created by computer scientist Fei-Fei Li, a dataset of 14 million photographs manually annotated, sorted and organised according to a taxonomy used to train algorithms. Malevé took significant care in investigating the invisible labour required to produce such annotations at such a scale. He based a series of re-experiments on the model of annotation used in Fei-Fei Li's original 1999 CalTech experiment, which required annotators to look at images at speed, a model which was subsequently formalised as the model of annotation used in ImageNet, according to the taxonomy of the WordNet thesaurus: 'To build ImageNet a large population of precarious workers were necessary to label and classify millions of images culled from the internet a task which had to be performed at speed and for which they are paid a few cents'. This, Malevé concludes, is a new economy of sight. Malevé's research is important in tracing how value categories, derived from the representational codes of photography and the taxonomies which structure linguistic classification, become mobilised in computer models of vision and importantly shows the limits and flaws of such models. Malevé concludes 'On the Data Set's Ruins' by pointing to an alternative direction of research; 'Instead of denying the collaboration, they should address the trans-epistemic dimension of the work carried out by the annotators and engage with it. This is hard work, because it questions the discipline, where it hurts most: at the level of its economy' (2020).

The computational, operative and non-human definitions of the image have been reached primarily through a concern with the automaticity of technological development, with the apparatuses of vision and how they are selectively deployed. In one important sense, this is the image beyond the human, cut loose, with a mind of its own, a non-human vision bypassing human meaning and in which human judgement has been removed, abstracted, encoded and instrumentalised. As Farocki and Paglen have pointed out, this is the image concealed from human view and presence of mind, which then needs deliberate acts of unconcealment to be returned for public scrutiny. Clearly such computational capability in the hands of the military–industrial complex remains a real and urgent matter of concern. At the same time the non-human dimension of computational technology has been seen in the cyber-feminist[10] and post-human context of the human/machine exchange as a matter of care. This is what Joanna Zylinska is striving for in non-human photography, to see in automaticity a positive reaffirmation of nature and a path of repair. In another important emphasis, automaticity, as Nicolas Malevé has shown in the case of computer vision, is all too human and the effort at unconcealment in his work has been precisely to reveal automaticity as a function of MTurk, Amazon's reserve army of labour who perform the repetitive tasks necessary to produce sufficient data for the algorithm to perform averaging tasks. The transitions from the digital image to the computational image are characterised by increasing complexity in which the image is part of larger systems of informational relay. In the image's relationship to transmission systems and large-scale datasets we arrive at the threshold of the networked image.

The Ecology of the Image after Photography

The term network and the network image encapsulate many elements of the socio-technical assemblage in terms of understanding the image as a relational field. The network image can now be understood to be constituted on a material base of computing in which the digital and algorithmic are the operational code. The network image overlaps with, but is distinct from, the terms operational and non-human, which are seen along with machine vision as specific deployments and applications using computational systems. In this concluding section the image is characterised as a property of the network of networked computers and the communication protocol of the World Wide Web. From its origins as a peer-to-peer, user-led network of networked computers, the Internet has developed into a corporate monster, dominated by closed proprietary operational systems. The dream of an open-source medium has landed in the algorithmic filter bubble, the Internet void. The early promise of the Internet, envisioned as a communication medium which did not have to replicate existing gatekeeping and hierarchical access to knowledge but could open up new freedoms of thought, has been seriously challenged and a new mood of pessimism exists about the Internet's future. The Internet is now shaped by the interests of big business and militarisation in different forms of surveillance and control – what Jonathan Beller has described as computational capitalism, James Bridle defines as 'the new dark ages' and what Bernard Stiegler has termed the recursive society and algorithmic governance. The argument now is about how much the user colludes, unwittingly or not, with the interests of data markets and what forms of resistance and alternatives are possible. Is there still a possibility in the global reach and technical resources of the Internet for new collectives and ways of addressing common problems at both local and planetary scales? Such questions are raised by works considered so far. Benjamin H. Bratton's *The Stack: On Software and Sovereignty* (2015) offers a compelling macro analysis in his conceptualisation of planetary-scale computing. From an interdisciplinary design perspective, he analyses the network of networked computers as 'a stack', an accidental megastructure, connecting layers, conceptualised as earth, cloud, city, address, interface and user, acting together in the stack as a sovereign state. Alternatively, Ingrid Hoelzl and Remi Marie, in *Softimage: Towards a New Theory of the Digital Image* (2015), offer a micro approach, analysing the shift from the geometry of photographic representation to algorithmic functions of the digital image, opening out into an explanation of the (photographic) image as a continuous actualisation of network data in an urban data space. Both Bratton and Hoezl and Marie's studies are accounts at different scales of the datafication of knowledge and value in human transactions, which bear upon the question of what is happening to the visual and value in culture. Each approach is absolutely necessary and at their best micro and macro analyses connect with the urgencies of both everyday and planetary-scale problems. The rapid and exponential scaling up of images, the tsunami of images uploaded and shared on the Internet, made possible by the ever-greater scaling up and refinement of the Internet's infrastructure, is currently the primary manifestation of the networked image, approaching the threshold of a medium. However, in regarding the networked image as a socio-technical assemblage, the networked image is also an embodied affect, working not simply as the remediation of a representational image, but as a mediation in life (Kember and Zylinska 2012).

In Hito Steyerl's essay 'Too Much World: Is the Internet Dead?' (2017), she points out that the Internet is more potent and extensive than ever, but at the same time it has started moving offline. What does she mean by this? Steyerl uses television broadcast as the inaugural historical moment in which images no longer record and document, but become active catalysts of events. Her specific example is the invasion of TV studios by protesters during the Romanian uprising in 1989 to make history, as she puts it. Since then images have become neither objective nor subjective representations of a given reality, but rather nodes of energy and matter, shaping and affecting the social world. As Steyerl puts it, 'Around 1989, television images started walking through screens, right into reality' (2017, p. 144). The Internet settled this realignment between image and reality for good. In many respects, Steyerl continues Baudrillard's vision of hyper-reality and the implosion of representation and she does so in the same terms as he did, by starting with the prototypical device of the era of simulation, that of the television screen. So there we have it; the image has only a residual relation to photography, held by a name. Now the focus of interest must be in trying to understand how the network image operates as a system of reproduction of social relations and subjectivities.

Gathering Understandings and Terms around the Network Image

Across all of the attempts to define the essential characteristics of the network image lies a recognition of the provisional nature of the language used. It is not a matter of settling on specific terms, but rather, in the accelerating logic of information, reaching for new understandings. There is a strong sense in the humanities of running to keep pace with events. The term network image, as has been said elsewhere, operates as a linguistic placeholder for something that is still unfolding. But right now, ask the network of networked computers, more familiarly the Google search engine, to provide you with facts about itself and the data it handles. The answers scroll through page rankings of results, which are hard to measure at a human scale. At the time of writing, various tech and media business websites tell us that currently 2.5 quintillion bytes of data are created each day, and the estimate now is that 1.7 MB of data are created every second for every person on earth and that 90 per cent of all data has been produced in the last two years. In order to maintain such exponential amounts of data the Internet currently uses 10 per cent of the world's electricity generation. In addition, more than 3.7 billion humans currently use the Internet. Google Search, used to amass these somewhat redundant facts, processes 3.5 billion searches a day and more than half of all web searches are made on smart phones. The point to be held in mind is that the visual image is produced by and works in and for data-driven networks. You could say the image is the network and the network is the image. Every minute of the day, Snapchat users share over half a million images and YouTube users watch over 4 million videos; 50 billion images have been shared on Instagram and estimates for how many image files exist run into the trillions. Whilst these statistics are themselves only possible by virtue of computation, the cultural and social impact of such changes over a relatively short period of time is less calculable. Such numbers tell little about the ways people feel about images and the ways images make them feel about themselves. Studies of social media have reinforced earlier anxieties and spurred policy concerns about the harmful effects of too much time spent online, especially for young people.

The ubiquity of the image (Hand 2012) has been used to breathe new life into photography, in arguments that the greater scale of image circulation denotes the arrival of the 'social photo,' which increases the human experience of the social world in a new form of documentary consciousness in which the world is consumed as an image, enabled by technology. Jurgenson (2019) argues that the social photo is a technological mediation and augmentation of life which articulates the self and sociality, rather than inherently diminishing or destroying them. It is here that the yawning gap is once more revealed between the image and data, the gap between what the search engine does and what it says, between how the image is perceived and how it functions. Defined as scalable data it is possible to consider the production and viewing of the network image as a training exercise for neural networks on the road to greater automaticity, but defined as photography, it can be seen as an amplification of the traditional role of social documentation and interaction.

The idea of the photographic image, expressed in network terms, has travelled a long way from any discrete representational definition of an image, coming closer to Baudrillard's definition of the era of hyper-reality, with a subject twitching in front of a screen at an immeasurable distance from a point of origin (1983). Rubinstein and Sluis's compound of representational and viral logics, pulling the subject in opposite directions, as well as the binary logic of the visible and invisible in Paglen's view of the weaponised destination of the operative image, are only two of the paradoxes that abound in the world of nonrepresentation. Hoelzl and Marie resolve the persistence of a representational logic in terms of defining the screen image a 'lure,' yet another kind of simulation, for the interests of data. In doing this they also retain the bifurcation of image as data and image as representation, only now thought of in terms of the expanded or hybrid photographic image.

Interest in the social boundary condition of the network image is central to Ingrid Hoelzl and Remi Marie's (2015) study of what they term the 'softimage,' one that combines expanded, operative and algorithmic dimensions. In outlining the shift in image production from print to screen, and from analogue to digital, Hoelzl and Marie identify the desire for endlessness in the image, a condition contained in montage, collage and animation, reinforcing the idea of the expanded image. At the same time Hoelzl and Marie pursue an understanding of the algorithmic image, no longer governed by geometric projection, but by computer processing. This they describe as a historic paradigm shift from projection to processing, from the instrumentalisation of the gaze to the algorithmisation of operations in which the image is a continuous actualisation of networked data. For the trajectory of the image being traced in this chapter, and in plotting the transitions across the digital to computational image, Hoelzl and Marie identify a number of key and paradoxical understandings of the new condition of the visual image. First, they make clear that the photographic image has undergone a technical transformation by virtue of the widespread adoption of the JPEG and the Joint Photographic Experts Group's definition of the JPEG as a category of images based upon bitmapped processing, rather than a mode of (photographic) image recording. However, having said there is nothing photographic left in the digital image, other than light as a source or value, Hoelzl and Marie go on to point out that it would seem that the photographic paradigm remains intact on the level of visual perception. They reach the same conclusion as the starting point of this enquiry; 'While the digital revolution erodes both the technique (geometry, projection) and the philosophy (transparency, truth) that underpinned it, the photographic paradigm seems to remain intact on the level of visual perception, so that today the photographic image occupies the entire field of representation as well as the one of vision' (2015, p. 3). But more important than their recognition of the continuing paradox of the photographic image, or algorithmic photography's Janus-face, as Rubinstein and Sluis have it, Hoelzl and Marie offer a new definition of the relational image.

> The image is the continuous actualization of networked data, and as such actively partakes in the daily configuration of what we call the urban data-space, the city considered as an ensemble of physical and digital data, of bodies and signals that communicate and commute via wireless networks and mobile devices. (2015, p. 3)

This is an important foundational definition of the networked image, which stresses that the image is no longer fixed or passive, that it is multiplatform and has a signaletic temporality by virtue of digital screening, compression and transfer. Further, Hoelzl and Marie stress that the screen is the current form of the image, which is no longer tied to a limited range of fixed locations. Their analysis of the new conditions of the image lead them to consider the image's relationship to augmentation, pushing the boundary of the image's relationship to space in defining urban space, the space of the city, as data-space composed of physical and digital data. This leads them to conclude that:

> [i]n this urban data-space, the screen, as a local access point to the networks, coincides with the image as the visible part of this data exchange, and the image-screen is defined in terms of its temporality, that is, in terms of the speed of network access and data transfer. The image, then, is nothing but the moment of network access. (2015, p. 126)

These are advanced positions in thinking about the image after photography, but in one way or another they continue to accept the necessity in the network of a representational logic founded upon photography. Of course, the image still communicates, speaks and performs to humans and it continues to do so through a cultural code of representation, thus apparently repeating the paradox of a nonrepresentational system whose screen image is representation. But the representational form of the computational image is no longer limited to the photographic, because software can draw upon any calculable graphic source and conventions, and in the case of the simulated photograph increasingly the software deploys montage, animation and painting techniques. Moreover, the immanence of the image as mediation operates in relationship to a host of embodied and transactional languages. As all of the authors considered here point out, the network image is technically and socially computational and relational. They point to the excessiveness of the image in its flows of graphical, textual, temporal and networked information. Why then is it necessary to retain the representational logic of the photograph to engage with the persistence of representation itself? Shouldn't the focus of attention be upon the very excess and overflowing of information they identify? The discrete charm of the photographic image can only be an obstacle to the new image programme and its visual characteristics. The continuation of the rhetoric of photography can only conceal the reality of the system of value which produces the image. In Bifo Berardi's brief introduction to Hito Steyerl's publication *The Wretched of the Screen* (2012), he makes the claim that 'history has been replaced by the endless flowing recombination of fragmentary images (p. 10). This is not only a reference to the image crossing the screen, but also a reference to the break-up of the spatial world ordered by Renaissance perspective and formalised in the focal plane of the camera. As Steyerl polemically puts it, 'The tyranny of the photographic lens, cursed by the promise of its indexical relation to reality, has given way to hyper-real representations – not of space as it is, but of space as we can make it – for better or worse' (2012, p. 26).

The network image is now the default of reproduction and hence not a tool in any simple sense but rather a complex cultural ecology in which the human/machine relationship is ever more closely entwined. In wide sociological terms, the applications and platforms of online communication can and have been used for progressive purposes, as pointed to by Manuel Castells (2000). However, the Internet, as Geert Lovink (2011) argues, is a network without a cause and the early promise of a radically empowering new and open democratic medium is now a corporatised environment, dominated by a few very powerful global tech companies. Lovink is expressing a new set of political problems from the position of a much longer critique of the organisation and control of media and communication in capitalism.

As the long argument for forgetting photography moves towards its denouement the question of the politics of the image comes to the forefront of attention and is developed in the next chapter. But finally here, in mapping the field of the image after photography, one thing above all else needs to be clear, that photography is a residual organisation of the visual and has little if anything left to offer in penetrating the thick fog of reality. Again, as John May argues, digital photography is 'an oxymoronic conjunction in which the incommensurability of both terms is masked beneath their apparent similarity' (2019, p. 13). (More importantly he places this conflict within a continued use of language, in which photography persists as a collective mental habit. Photography is an historically limited consciousness, an echo of the past, which suits the corporate bureaucracies of the network to simulate. Photography exists to paper over the cracks in the real conditions of data extraction and contain the possibilities of an altogether more democratic future. The problem, therefore, for the continued academic investment in tethering photography to the new ontologies of the image is that it unwittingly abets the bureaucracies whilst labouring under the conceit of being critics. Continuing to draw upon the language and technical ontology of photography can only set back the time when new ways of seeing can be expressed and a new field of knowledge is framed. Thinking about the image after photography opens up the whole spectrum of visuality in media and across every platform; it looks at the screen image, not the discrete simulated image within the screen. We need to wake up, for time is finally up on photography, which henceforth can only be a conservative force, propping up an old order and concealing new lines of power and their potential opposition.

8

The Politics of the Image

To say that the condition of the future is technological in no way means that this condition is a solution: it means, on the contrary, that the condition is a problem (and not just a question), and that what is required is a 'great politics' of technology, which must become a 'great health,' that is transvaluation.

... even if, eventually, capitalism will disappear, the transformation currently underway is unlikely to lead to a straightforward exit from capitalism any time soon. In short, the questions and the problem are to ensure that we do not disappear along with capitalism.

Bernard Stiegler. 2019. *The Age of Disruption: Technology and Madness in Computational Capitalism*. Cambridge: Polity, pp. 293, 295.

The logic of forgetting photography looks to understand photography's institutionalisation across the twentieth century, which was essentially a conservative process, as its passage into the museum and the academy demonstrates. Rather, in the manner of the British monarchy, paradoxically redundant but still functioning, photography remains a constituent element of the social relations of the mode of reproduction. Democracy would certainly be advanced by the abolition of the monarchy, just as much as forgetting photography would quicken wider understanding of the new operations of the image. Reproduction maintains an existing social order, even in crisis, upon which the mode of production depends.[1] As the constituent mode of twentieth-century analogue reproduction photography was not only a technical apparatus, but also a set of social relations, a 'way of seeing' in fact. Today reproduction is based upon computational networks, as we have seen, in which a new image and with it new ways of seeing and acting are emerging, alongside the remnants of the old. One of the problems for image studies, especially those emanating from the cultural and media field, has been the tendency to privilege either technology or the socio-economic order as the primary logic of explanation. Now, however, there is a general consensus in media research that technology is cultural and culture is technological, with a new emphasis upon the complex entanglements of human and machine interactions. The image is now thought of as a socio-technical assemblage. With computational reproduction the cultural and technical are more enmeshed, making it harder to see the mode of reproduction as a distinct system. This study has been more than aware of theoretical perspectives which start from entanglements as a means to trace the active micro agents (of power) in affective networks, rather than assume a knowledge of a given social or technological order. This was the burden of Latour's work on reassembling the social[2] and many detailed studies in science and technology that followed. Entanglements are also central in post-human thinking and are conceptually crucial to Karen Michelle Barad's ideas of agential realism. Analyses based upon conceptual models of entanglement argue that the world is not a given, but has to be traced as it emerges or is constituted through 'intra-actions' and 'networks of association' between humans and their/other objects. Relational analysis is proposed as a better way to understand the manifold problems of the way humans organise things and points the way to alternative organisation and decision making, which could be applied to understanding the current crisis of the image. However, the networks in which academic analysis gains agency are not seamlessly connected and extended to the networks of governance, but rather selectively filtered, and the dissemination of knowledge is slow, diffuse and indirect. The protected networks of economic and political agency, by contrast, are fast, direct and with immediate consequences. Power is naked in its macro guise, nowhere more unambiguously exposed than at the point at which nation states engage in techno-wars and necropolitics (Mbembe 2019), just as much as profit drives continued financial investment in fossil fuel extraction or governments enact laws restricting workers' rights because they adhere to free-market ideologies.

The prevailing global mode of political economy remains the economic system of capitalism, based upon profit, accumulation and wage labour.[3] Since the fall of the Berlin Wall and the subsequent collapse of the USSR, neoliberalism, or free-market capitalism, has established itself across the globe. Over the period of modern history in which photography has been entangled it has participated in the catastrophic destructive events of the First and Second World Wars, the US atomic bomb attack on Hiroshima, the post-Cold War between NATO-aligned states and the USSR, the US war against North Vietnam, South African Apartheid, the USSR invasion of Afghanistan, the US invasion of Iraq and the subsequent, continuing 'War on Terror', to name only the most obvious from a Eurocentric perspective. The period also encompasses world economic growth driven by fossil fuels and the consequent climate crisis, neoliberalism's deregulation and monetarist policies that have created a global reserve army of cheap labour in which the top 1 per cent of the world's population currently own 44 per cent of its wealth. But why should these bare facts of violence and power be considered part of understanding what happened to photography and what is unfolding with the computational image? In the case of photography, while it would be argued that photography was an independent witness to the many human consequences of the exercise of power, it is also the case that it rarely, if at all, made power the subject of its images; rather, it took part in the exercise of power. Now, the computational image, born as a military strategy,[4] seamlessly and invisibly crosses the borders of civil and military communication and command. In matters of extant power, the fine grain of scholarly study meets its own existential crisis in the knowledge economy, whether reflexively recognised or not. Somehow, understanding the historic role of photography and, more importantly, understanding the new image's enlistment in evolving global systems of power requires a more engaged and aligned starting point. From the position taken in this account, the most insightful and illuminating writing about photography over the course of the twentieth century came from politically engaged thinking and activism, Walter Benjamin, Susan Sontag, Allan Sekula and John Berger being amongst the most widely read figures, writing from political traditions of independent intellectual work which cared about what happens in the world and which strove to combine understanding with action.

Politically Committed Photography

Twentieth-century socialist and democratic movements used photography in conjunction with print to agitate for both reform and revolution, largely through documentary realism, with some notable exceptions in which the radical potential of montage was realised.[5] The international worker photography movement, first appearing in Germany and the Soviet Union in the mid-1920s with the organisation of amateur photography groups and the publication of specialised magazines, encouraged workers to document their working lives. Worker photography was supported by the Comintern communist movement as contributing to the revolutionary transformation of the world, but by the outbreak of the Second World War had collapsed.[6] Terry Dennett and Jo Spence revived the ideas of the worker photography movement in establishing the Half Moon Photography Workshop in Britain in 1974 together with launching the broadsheet magazine *Camerawork*. After they both left Half Moon, they went on to publish two edited volumes of critical papers under the title *Photography/Politics: One* and *Two*.[7] Over the same period in Britain a network of independent community darkrooms and print shops emerged with the express purpose of countering mainstream media's dominant model of representation.[8] In the USA, Martha Rosler was notable in using photomontage to produce works against the US war in Vietnam, such as her 'House Beautiful: Bringing the War Home' (1967–1972),[9] as was Allan Sekula in his 'Untitled Slide Sequence' from 1972, a 35-mm slideshow that presents 25 still images of aerospace factory workers as they leave their shifts at a General Dynamics Convair Division factory in San Diego, where several of them likely helped produce the F-111 military planes that flew in Vietnam.[10] Also in the same period, Caroline Hunter, working for the Polaroid Corporation as a chemist, with her partner Ken Williams, founded the Polaroid Workers Revolutionary Movement, in order to campaign against Polaroid's support for Apartheid by indirectly supplying its ID-2 camera system to the South African regime. Designed to increase the flash for dark-skinned subjects and with two lenses which produced a portrait and profile on the same negative, the ID-2 camera was used to create images for the infamous passbook identity cards. No doubt there were many other examples of overtly political photographic practices over the period.

In retrospect the fault lines of photography as an oppositional medium were already in evidence through the growing theoretical challenge to photography's transparency and truth claim, upon which social documentary depended. Photography as an independent, radical, community-based practice in the UK had also declined by the end of the 1980s, through the loss of public funding; however, elements of the practice continued at points with the support of sympathetic elements in academia.[11] By the 1990s photography was enjoying a new international status within the art field and elements of independent critical practice were absorbed. But the revolution which swept away radical and community photography was the digital revolution, precisely because photography was suddenly everywhere, no longer documenting everyday life but part of it. The world of representation had been replaced, as if overnight, by the image as performance and the image as information. In this respect it is no longer possible to think of photography under its current conditions as a tool to be used against its financial masters, because it is no longer possible to think of photography as a unitary medium. Now the issue for socialist, progressive and critical thinking is how the network image operates in the balance between independence and privatisation and between commodity fetishism and social use. It is also a practical cultural issue of the contexts in which the networked image is met and how arts, media and educational institutions can critically engage with it.

The Politics of the Image under the Conditions of Neoliberal Capitalism

Over the last decade, capitalism has been qualified in a number of important ways, in order to identify more closely its centre of power and mode of operation. In the twenty-first century capitalism has been defined primarily as global and neoliberal, but with some important qualifiers. Achille Mbembe (2019) developed the concept of *necrocapitalism*, 'defined as contemporary forms of organizational accumulation that involve dispossession and the subjugation of life to the power of death',[12] in which neoliberalism's exclusions of surplus populations is put at the logical centre. Naomi Klein (2008) wrote of *disaster capitalism*, suggesting that neoliberalism deliberately exploits and engineers shocks, which act to divert citizen attention and justify questionable policies. Yann Moulier Boutang (2012) coined the term *cognitive capitalism* to explain the rise of the knowledge economy in which the notion of scarcity, which traditionally permeates economic analysis, refers less and less to physical goods and is applied to notions such as cognitive attention, time and affective attention. Jodi Dean (2005) defines the state of *communicative capitalism* as a form of 'ideology' in which 'values heralded as central to democracy take material form in networked communications technologies' (2005, p. 55) and instead of producing greater equality technology undermines and restricts social progress. Dean (2020) considers whether something worse than neoliberal capitalism is emerging in exploring the idea of *neo-feudalism*[13] to account for extreme inequality, precarity and monopoly power in which a property underclass will survive by servicing the needs of higher earners as a new class of servant, or worse, serf. Another related term denoting a qualitative change in capitalism is Colin Crouch's (2004) *post-democracy*, through which he detects a decline in the effectiveness of democracy precisely in those parts of the world in which it was most strongly established. Luis Suarez-Villa (2012) wrote of *technocapitalism*, exploring new forms of commodification and corporate organisation central to macro global social dynamics, in which advancement is science and technology associated with computing and telecommunications produce, rather than social inequalities.[14] Nick Srnicek (2017) defines the condition of *platform capitalism*, in which he argues that while smaller than industrial production, the digital economy is a pervasive infrastructure for the whole economy, whose collapse would be devasting. The position of the giant corporate tech companies is therefore becoming a hegemonic model. *Computational capitalism* is used by both Bernard Stiegler (2019) and Jonathan Beller (2018). For Stiegler, technology is driven by an instrumentalising technocracy, which he defines as the 'new barbarians,' and which dominates historical memory, leading to a loss of collective hope, which he deems a proletarianisation of the human mind. For Stiegler, technology needs to be rescued and reset. For Beller, computation has functionalised culture as part of the political economy, such that culture is now not the sphere of the recuperation of labour, but an extension of the factory shop-floor.[15]

Developing Progressive Political Cultural Strategies and Tactics

In what ways can these political and economic perspectives on the structures of contemporary global capitalism be related to an understanding of the new conditions of the networked image? The overriding common denominator in the majority of the accounts is the increasingly central place of technology and communication media in the political economy, or, more specifically, the structural economy of computational networked telecommunications, 'the accidental megastructure' as networked computing is conceptualised in Benjamin H. Bratton's (2015) book *The Stack*. Computational technology is of a planetary networked scale, and although not considered as over determined, nevertheless most accounts consider that technology is a driver towards a state of social, economic and material automation, organised around optimisation, visibility and accountability. The image is at the forefront of this future-orientated technological scenario in the development and deployment of machine vision technologies. Such developments have given cause for alarm, most obviously about even greater unaccountable incursions of surveillance, but also about 'outsourcing' decision making to autonomous machines, using the predictive capabilities of algorithms. James Bridle (2018) gives a compelling account of technology's complicity in the crises engendered by neoliberal capitalism, sounding a stark warning of the dangers of seeing technologies as autonomous functions and neutral tools. What is needed in the 'new dark age', Bridle describes, is a new public technological literacy in order to take back control and determine the future. From the perspective of forgetting photography an important part of that task will be developing a new literacy about the image after photography.

The Culture of Hypermodernity and Working with the Paradoxical Present

The cultural and individual consequences of the combination of neoliberal economics and informational technologies are also necessary to consider in the new image culture, the everyday world in which we generate and interact with networked images. Computation continues to reorganise affluent social life by seemingly making life easier and things more readily available, but at the same time subordinating individuals to the instrumental flow and optimisation of calculation. This, it seems, is the central paradox of current times, of what Gilles Lipovetsky with Sebastien Charles (2005) describes as hypermodern times, an analysis which articulates the shrinking of any confident view of the future, creating the condition of the paradoxical present. The defining social and cultural figure of this accelerated modernity and the paradoxical present is the paradoxical individual, for whom order and disorder, moderation and excess, independence and dependence, regulation and deregulation are conjoined reflexes in a consciousness of conflicting actions, desires, pressures and times.

Like many of the critics of capitalism cited above, Lipovetsky and Charles recognise that capitalism has moved on; reconfigured by global liberalisation and deregulation, it has reshaped itself as a consummated modernity, a modernisation of the modern or, rather, as an accelerated hypermodernity. The consequences of hypermodernity's crisis of the future are seen in the exhaustion of ideology and the stalling of democratic politics, replaced by narratives and projects of science(-fiction) and technology. However, any future embraced by science and technology is a risk, insofar as it could makes things worse or it might make things better. It is therefore a gamble rather than a certainty. Whilst the present remains the future's horizon, chained to capitalism's short-termism, the present also expresses anxiety about future sustainability and survival. Awareness of environmental catastrophe has become a permanent feature of the present, which is changing behaviours and exerting social pressure. Meanwhile, science and technology increasingly model future damage at national and planetary level, aimed at evaluating and overcoming the dangers identified and avoiding future risks. As Lipovetsky remarks, 'Hypermodernity has not replaced faith in progress by despair and nihilism, but by an unstable, fluctuating confidence that varies with events and circumstances' (2005, p. 45).

Conflict, Time and Chrono-Reflexivity

The readjustments to both the past and future are, as Lipovetsky outlines, ordered by the primacy of the here and now, creating an unprecedented social temporality, marked by the deregulation and individualisation of time. Time has shifted from the future to the present, producing the effect of shrinking time's horizon. Time is in conflict with itself as the world of consumption, enforced leisure and the free labour of the Internet continues to deregulate the collective temporal constraints of the institutional and organisational world, and, 'as a result individual activities, rhythms and itineraries have become de-synchronized' (2005, p. 51).

Time is experienced paradoxically. Either there is too much time or not enough time. Depending on your position you can be time-rich and financially poor or the reverse. Time is experienced as a series of different qualities in competition with each other and its management has produced chrono-reflexivity. The networked image is a chrono-reflexive agent, the perfect moderator of time against time, a vector of the paradoxical present, providing a constant stream of visualised data against and with which the individual can measure their distance or proximity to others, providing a virtualisation of where they are, who they are connected to and ultimately who they are. But it is the virtualisation of what is taken here as the real which needs examination. A new literacy, a metalanguage for technology, as Bridle puts it, is a means of investigating what Baudrillard spoke of as the 'satellisation of the real' (1983), as much as it would be a means of reanimating and reembodying reality through progressive change. As Nick Srnicek and Alex Williams (2015) have observed. 'The utopian potentials inherent in twenty-first-century technology cannot remain bound to a parochial capitalist imagination; they must be liberated by an ambitious left alternative' (2015, p. 3). But, as Steigler points out, this is a task which is much easier said than done and one which marks a theoretical return to the very grand narratives from which progressive intellectual work has been trying to escape (2019, p. 293). Stiegler's observation doesn't make the attempt at a social collectivist programme any less reasonable, but points out that a successful formula for change still remains to be articulated.

The Utility of Thinking Photography after Capitalism

Ben Burbridge's *Photography After Capitalism* (2020) is considered here as a companion to 'forgetting photography' in its critical examination of the new conditions of the image, although Burbridge might disagree given that he retains the idea and term of photography as culturally useful. His book takes up a number of the themes of the crises produced by the contemporary form of capitalism, seeing photography as part of a political and technical economy. Burbridge focuses upon photography not as representation, but as labour, drawing in particular upon the ideas of *cognitive* and *communicative* capitalism. He argues that exclusive attention upon representation masks photography's direct and immediate relationships to a neoliberal economy. Burbridge sets out to unmask photography's economic position, asking what digital image labour does, for those employed as operatives, as well as how it affects the rest of us as unpaid users in the network. He notes the asymmetry defining the field of labour between low-paid, 'unskilled,' precarious workers and the relative privilege of cognitive labour. But he also notes a deeper logic of *communicative capitalism* when he says, 'Society has not been liberated from toil, so much as work has been repackaged as freedom. Many of us are continually haunted by the repressed suspicion that every part of our existence – the things we love, the relationships we forge, the beliefs we profess to hold – are, at some level, always already co-opted by the imperatives of profit' (2020, p. 135). Burbridge locates photography within the informational economy, noting its contradictory mediating position between science, journalism and art, but he also sees progressive, educational and potentially emancipatory dimensions of image practices. His method of illuminating this is twofold. As a trained art historian, he gives as examples the work of artists using video and photography, citing in particular the work of Andrew Norman Wilson,[16] which addresses the subject of digital image labour. In this he sees a productive space for art to produce speculative images of alternative futures, beyond what neoliberalism currently holds, which he argues impact in the real world by offering concrete models of better ways of life. He also sees art as being able in certain circumstances to perform or map the power interfaces of capitalism in that, as an imaging making machine, photography provides a way to situate mechanical and cognitive labour's relationship. Burbridge's second method is to reach for a radically expanded conception of photography as labour beyond capitalism, arguing that photography offers the space for collective subjectivities to develop. Drawing upon the different analyses of post-capitalism made by Paul Mason, Nick Srnicek and Alex Wilson, together with the work of J.K. Gibson-Graham (2006). Burbridge sees photography operating as a means of sharing in a post-work world. Post-capitalist thinkers are by no means aligned in their political strategies, but in envisaging the end of capitalism they open up the possibilities for building a post-capitalist world in the present.

In locating photography as part of a potential post-capitalist strategy, Burbridge recognises both the limits and possibilities, if not the paradoxical position, of photography, where, on the one hand, it has close affinities with the individualistic logic of neoliberalism, as well as being the exploitative labour of the image world factory, but, on the other hand, 'the operations of [the image world] can be sabotaged, different forms of ownership developed, the value it produces reclaimed, redistributed and shared' (2020, p. 147). Burbridge argues that photography is an accessible 'tool' of democracy, while recognising that with networked computation the field of visual communication has been extended, paradoxically by means of a concentration of the ownership of the technical means of production and the untaxed profits it produces.

In the argument for forgetting photography, the title of Burbridge's book could easily be reversed to consider capitalism after photography, which in many ways his book does by considering the realities of the new forms of labour of the networked image. Photography was industrial capitalism's image, operating within the reproduction of the social order of the mode of manufacturing production. Burbridge emphasises the continuity between film and photography in industrial capitalism and the computational image of post-industrialism, through the labour of those who manufacture images. The comparison of Harun Farocki and Norman Wilson's artworks, charting an historical path of workers leaving the Lumière brothers' film studios, linked to Kodak workers and then to the low-paid scan operatives of the Googleplex, is meant to link the analogue and the digital through labour. But the comparison of workers leaving factory gates masks what is going on inside the Googleplex factory, which marks a distinct break in the technical system in which the image is manufactured. Photography was part of industrial capitalism's world picture and hence why it came to be analysed within the Marxist critical tradition as an ideological function of representation, and why cultural activists called for an alternative and oppositional photography. But the ideological function of the photographic image as representation did not account completely for photography's reproductive role in maintaining the social relations necessary to industrial capitalism. Photography's imbrication in and across industrial capitalist reproduction is evidenced by the very taxonomies of photography, which correspond exactly, with rare exception, to the social mode of life necessary to the mode of production, hence as one example family and wedding photography. It was not photography's realism but photography's naturalism which served the hegemony of twentieth-century capital. In contrast the computational image's relationship to the social relations of reproduction is only now emerging with the computational mode of production.

Recourse to the Autonomy of Art and the Problem of the Art Field

In Ben Burbridge's *Photography after Capitalism* (2020), the most compelling examples of works that reveal something of the image's relationship to power reside in the field of art practice. It might seem obvious to state, but art whose subject is digital media is most often made with digital media, employed across a spectrum of tasks from research to the subsequent processes of converting and importing analogue material, through to composing, editing and grading digital media for digital display. These are works made about image technology, but in which the technology used is not wholly present for inspection in the work. Digital media is used as a set of tools to combine and compose new image works which can replicate the aesthetic of previous media. Claire Bishop's point about a lack of recognition of the dependency upon digital media in media artworks displayed in the art museum comes back into focus here, in a situation in which the digital condition of a work, whose subject is the digital, is masked by the very media it uses. Julian Stallabrass (2010) makes a parallel point in discussing the works of Jeff Wall. He notes that while Wall's works requires the use of digital media, there is little discussion of the effects of digitisation in critical literature on Wall's work. In response, Stallabrass argues that the digital photograph is a new medium and should be addressed as such. If it is not, then as presented in Wall's gallery and museum installations something is being concealed and the viewer is presented with 'a state of half-photography in which each surface has been digitally brushed over and bent to the will of the artist' (2010, p. 106). This is in no way to dismiss the import of such works cited by Burbridge, but rather to problematise the cultural form and institutional contexts in which such works circulate and are given meaning.

The art field has its own institutional critique (Fraser 2005), as well as an established association with critical art practice focused upon technology, the latter having gathered more interest and support in art biennales and major gallery and museum exhibitions over the last decade. There is an art market for collectable art works made with and about technologies, variously categorised as net art, post-Internet art, digital art, contemporary/conceptual art, political art and installation, in which works are acquired for both public and private collections.[17] For example, Ian Cheng's 'Emissaries' trilogy of graphic computer animation (2015–2017) exhibited at the Serpentine Galleries in 2018 had already been acquired by MoMA in New York. Artists whose work was discussed by Burbridge because of their critical take on imaging technology, such as Hito Steyerl, Adam Broomberg and Oliver Chanarin, Trevor Paglen, Jon Rafman, Joachim Schmid, Amalia Ulman, Mishka Henner and Andrew Norman Wilson, all have established dealer representation, in some cases with blue-chip designation. As Burbridge himself says in discussing Steve McQueen's 2007 film 'Gravesend,' which looks at the exploitative labour involved in coltan extraction and processing and was sold as an edition of six,[18] 'Art, like photography, is a two-faced janus. What is critiqued at the level of content is reproduced by its operationalization in the art world on which its visibility as critique relies.' Or, as Hito Steyerl remarks, the 'politics of art is the blind spot of much contemporary political art' (Steyerl 2010).[19] Artists need to be represented and need to make a living and the art market's relationship with the art field is a main filter by which works comes to be seen as well as to be considered culturally significant. Lucy Sollitt's 2019 report 'The Future of the Art Market,' published in partnership with Arteia, Arts Council England, Creative Scotland and DACS, commented on the likely future of digital art. The executive summary gave an upbeat message about the future of digital art, suggesting that the market will grow and that the prevailing art market view that it is a hard-to-sell niche will be overcome with shifts in production, distribution and preservation of digital-born artworks.[20]

Curators and academics have also invested in shaping a distinct 'genre' of art made with and about technology, ensuring its place in the art historical canon, in museum collections and in the European/US and now global art market. Historically, art made with and about technology has had a number of mutating subtitles reflecting technological development, including machine, electronic, computer, interactive, digital, net and new media. Transmediale, founded in Berlin 1989, started life as an art-based video and film festival promoting new cinema, which has grown into an international organisation promoting art and digital culture. ISEA, formerly the International Society for the Electronic Arts, founded in Utrecht in 1990, started from a symposia on electronic arts. Christiane Paul, who published *Digital Art* in 2003 charting the rise of digital art in the context of the art museum, was also responsible for establishing 'artport' at the Whitney Museum in 2001, an online gallery space for net art and new media art, providing access to original art works commissioned specifically for artport by the Whitney. The historical canon of art made with and about machine technology remains open and contested, because the acceleration of computational culture makes attempts to forge a linear path difficult and because the 'contemporary' of art is now also contested. One relatively recent attempt to assemble a canon of art made with and about technology was the Whitechapel Gallery's 2016 exhibition, *Electronic Superhighway (2016–1966)*, curated by Omar Kholeif, which included 50 artists and 100 works. Framed by the term computer art, Kholeif established a reverse chronology in which contemporary works made with digital technology focusing upon digital technology and the Internet were connected with early pioneers of video art such as Nam June Paik and works from the 1966 Experiment in Art and Technology (E.A.T.) group.

In the repeated recourse to citing art as the arena of technology's revealing, the question arises of what the 'media' object under scrutiny is. The parsing of academic knowledge according to ever more tightly drawn boundaries and subfields, in this case to establish a canon for art and technology, works largely to depoliticise media outside of the art field, whereas a politics requires the joining up of knowledge of specific actors with the sphere of everyday operations. Gaining an overview of the relations within any media practice is consistent with the call by many of the writers considered across this study for a metalanguage of computational capitalism.

There is a singular irony about the current institutionalisation of a specific category of political media within the art field. Historically, groups of cultural activists decamped from the analogue institutions of the art field to the Internet, seeing in its earlier formation a new space of freedom unencumbered by institutional gatekeeping and the elitist trappings of art. The Internet was championed as a radical new space to explore cultural value and to create new open and democratic collective actions. The open-source software movement along with tactical media and hacktivism attempted to direct the Internet away from its growing proprietorial and finally corporate form. Net artist Mark Tribe founded Rhizome in 1996, a not-for-profit organisation in New York, and the ArtBase, an open repository for work made on and about the Internet, was set up in 1999. Rhizome became affiliated with the New Museum in 2003. In London, Ruth Catlow and Marc Garrett founded the arts organisation Furtherfield in 1996 with the expressed intention of engaging Internet users in collective methods of working, rather than the exclusive focus upon the individual artist, creating software platforms for engaging users in shared creativity exchange. In 2004, Furtherfield opened a physical gallery in North London and since that time it has received public arts funding. Rhizome and Furtherfield now collaborate with established universities and museums interested in the methods and open principles of independent arts organisations, at a point in time when the independence of established institutions is threatened.

The art field has made an important contribution to the development of a new set of critical understandings of the computational image within the discourse of both contemporary art and Internet culture. However, the computational image as the default of reproduction encompasses the whole mode of production and social life, proliferating hybrids of nature and culture, human and machine in all fields of knowledge/practice. In this respect the art field, which 'purifies' hybrids in order to define its own object, cannot be expected nor is able to produce the kind of 'metalanguage', or the 'new literacy', demanded by the new conditions of computation. In political terms the global art market and its collecting institutions continually recuperate and contain any potential for radical cultural and institutional change, precisely through ownership and collection. One of the original aims of early net art was to operate outside of the institutions of art and seek new communities and audiences, an aim recuperated by the commodification and uneasily negotiated by independent cultural organisations.

Public Cultural Institutions and Digital Development

Understanding the impact of computation will require every type of institution to recognise and begin to engage critically, not only with the knowledge skills of digital marketing and web content management, but with the cultural and ultimately political questions of value posed by the network and computation. There is a growing recognition in research communities across science and art that such a task is necessary and will require new knowledge as well as crossing different knowledge boundaries to succeed. In the field of cultural research, transdisciplinary approaches and collaboration across institutions are increasingly supported. Of specific interest in considering the image after photography is the example of the collaboration between the Photographers' Gallery (TPG) and the Centre for the Study of the Networked Image (CSNI).[21]

In giving an account of the Digital Programme of TPG, potential critics of the call to forget photography will note the glaring irony that it is an institution dedicated to promoting wide public enjoyment and understanding of the medium of photography and the work of photographers, which is prepared to open its door to the medium that is in the process of consigning photography to history. On the other hand, what better context to consider the conditions of the computational image and the ways of seeing it is bringing forth than in an institution which for the last 50 years has dedicated itself to the medium in which a modern consciousness of ways of seeing first appeared? Moreover, one of the aims of TPG is 'To be the driving force for debate and new thinking about the role of the photographic image in society today', suggesting that its programme is open for change. This was evident in what the current director Brett Rogers had to say when interviewed by *The Guardian* in 2011.

> Photography has become a very natural, even compulsive thing with the coming of the mobile phone camera and relatively cheap, hi-tech digital compacts. The democratisation of photography and distribution of photos via social networks has changed everything, and we, as curators, cannot simply stand back and ignore that.[22]

TPG was founded in London in 1971 by Sue Davies, its first director, who established its original home at 8 Great Newport Street in London's Covent Garden. It was the first gallery in the UK offering a dedicated space for photography and photographers. In 1980, the Gallery acquired a neighbouring premise at 5 Great Newport Street, extending its exhibition spaces and providing room for a bookshop and café. It was also able to accommodate an area for print sales. In May 2012, TPG reopened after

24 months of refurbishment to its building in Ramillies Street, now with the addition of two additional floors designed by the architectural partnership O'Donnell + Tuomey. The cost of the refurbishment was largely funded by a £3.6-million grant from Arts Council England's Lottery Fund, £2.4 million from the sale proceeds of TPG's previous building at Great Newport Street and £2.8 million from foundations, trusts, individuals, corporates, an auction of donated photographs held in 2011 and other public funds. TPG has had three distinguished directors over its 50-year history. Sue Davies was in post until 1991, succeeded by Paul Wombell, who had moved from directing the Impressions Gallery in York. Wombell was succeeded in 2005 by Brett Rogers, the current post-holder. Across its half-century the gallery has remained remarkably consistent in its singular dedication to the photographic medium. TPG is a registered charity and is part-funded by Arts Council England as well as support from individuals, companies, trusts and foundations with profits from its bookshop, print sales and gallery contributing to the exhibition and public programme.

The Digital Programme of TPG was inaugurated with the appointment of Katrina Sluis as its first, part-time curator in 2011 and the installation of the Media Wall and computer backend, mounted in a prominent position in the entrance to the refurbished building. The context for this development can be explained by a confluence of factors, as Brett Rogers recognised, when she initially described the installation of the screens as 'the wall for all', in an acknowledgement of the new ubiquity of the image. The new mobile condition of the image and its circulation in social media connected with the mission of TPG to promote the enjoyment and understanding of photography for all. For some considerable time prior to the TPG initiative, the impact of the 'digital' in culture had been the subject of UK national policy initiatives at the Department of Culture, Media and Sport (DCMS) and the Arts Councils. Digital arts policy was being formulated, as it always has been, around the interests of economic competition and it was the needs of the digital economy which directed funding initiatives for arts organisations. The National Endowment for Science, Technology and Art (NESTA) was created in 1998, with £250 million of National Lottery funding (later supplemented, in 2006, with a further £75 million of National Lottery funding drawn down over five years). The initial concerns of NESTA were directed at what was seen as Britain's technology innovation 'gap' and how to promote innovation, research and development to close it. Similarly, Arts Council England (ACE) saw the technology issue for the arts as a catch-up exercise, where the emphasis fell upon training and skills. ACE's recent Digital Culture Network expresses this perfectly in offering cultural institutions help in digital marketing, ecommerce, data analytics, search engine marketing and outsourcing policy and strategy to arms-length organisations.[23] Digital policy continues to be primarily directed towards a digital skills agenda, on the view that access and training are the key to future creativity, marketability and sustainability of the arts. In parallel, arts policy has also been struggling with dwindling audiences for the arts, and the recognition that audiences for the arts are not getting any younger, nor are they representative of the diversities of culture.

The twin policy and funding agendas of digital skills and audience diversity were brought together in the prevailing arts policy perspective upon marketing in which TPG, along with other arts organisations, was drawn to see the Internet primarily as a recruitment tool. One such argument at TPG was that young people were engaged in digital culture and that programmes of activity based upon the popularity of digital culture would encourage a new generation into an educational relationship with photography. This as it turned out was the basis for the funding rationale which led TPG to apply to the Esmée Fairbairn Foundation,[24] which initially funded the new Digital Programme at the gallery for two years from 2011. What was interesting in this educational framing of the digital was the emphasis upon the culture of the digital rather than the skills agenda, although the two are not mutually exclusive. It was, however, an open door to an investigation of the culture of networks, something photography curating had not considered.

Funding policy together with the new ubiquity of photography fed into the context of setting up the Digital Programme at TPG, but how did the Digital Programme relate to the context of curating photography? Curating photography remained bound by the modernist logic of the gallery apparatus of exhibition, which by the late 1990s oscillated between two competing aspects of modernist aesthetic formalism, one in art and the other in photography. For photography the modernist tradition turned upon the photograph as document, realised in the unique vision of the photographer and expressed by the primacy, if not sanctity, of the photographic chemical negative. Conventions of display were strictly developed on the basis of standard print sizes and frames, following the negative format; 35-mm negatives, for example, were displayed in 20 x 16-inch black frames on white or grey walls. The aesthetic norm was based upon black-and-white photography, held to be the original, from which colour photography was a derivative. This can still be seen in the aesthetic of TPG in which black-and-white was chosen as the iconic brand for the refurbished building. The aesthetic puritanism of photography curation gave way to, or intermingled with, the aesthetic modernism of art, exemplified by the Turner Prize display of Wolfgang Tilman's work at Tate in 2000. The change to a more expansive 'art curation' was also reflected at TPG, in the establishment of an annual photography prize whose origins lie in an earlier London Citibank sponsorship in 1992, which developed into the annual TPG photography prize launched in 1996, which from 2005 was sponsored by Deutsche Börse, awarding £30,000 to one of the four selected nominees for a body of work produced or exhibited in the previous year. In 2015, Deutsche Börse set up a photography foundation and in partnership with TPG the annual competition has since been known as the Deutsche Börse Photography Foundation Prize. Over the past two decades and more, the nominees and prize-winners have been drawn to a carefully curated form of cultural diplomacy, a balancing of risks, a modernisation of the modern, in the institutionalising and corporatising of the photographic canon, charting its way through the hypermodern anxieties and insecurities of the global present. Within the modernist gallery apparatus, photography curation reproduces an image of the contemporary world in order to secure the medium of photography and its essential link with external reality. What else can or could photography curators do, it might be asked.

In the tradition of modernist curating, the photograph operates as a semiotic, figural object, however formally narrated or styled, in which the viewer constructs meaning through the representational elements of the image assembly. In effect there remains a mediated route through the exhibitionary complex from the image in the gallery to the external world of events. But what of the image which escapes the exhibitionary complex, the image, which in Steyerl's writing (2009) has crossed the screen, a fugitive image as part of the real which pervades consciousness, no longer a representation of any external reality, but a transmission, an operation, a node in a network, a replaceable screen in which the fleeting image assembles? With modernist photography, the image, the specificity of its medium and the image's technical mode of display are stabilised and aligned by the institutional practices and conventions of viewing. In radical contrast, the network image is a relational object with an unsteady state, a set of continually changing relations, governed by the corporate institutions of computational network systems. This begs the question of what an image of the fugitive image would be and how it might be possible to consider curating the network image and what form it might take within the exhibitionary complex.

Curating the Computational Image in the Network

How to approach curating a cultural object which is defined by its relational and distributed field in an institution committed to the furtherance of photography as a defined historical and cultural object was the creative challenge Katrina Sluis set herself. Her programme of interventions over a seven-year period represent one of the most developed and sustained attempts to look at the new conditions of the image, to look at digital technology not simply as a tool of communication, but as a culture of value. In its institutional context the programme was framed as a curatorial enquiry into how technology was changing photography, but its larger ambition sought to build a public knowledge of computational culture as media and in particular how its platforms and interfaces operate. In this ambition it is possible to detect a continuity with modernist Left and critical cultural image practices, with its concern to develop forms which demonstrate media's links with power and their consequences. Katrina Sluis, who is also a founder co-director of CSNI based at London South Bank University, saw the position of curator of digital programmes at TPG in more expansive terms than programming the Media Wall. Seeing the obvious limits and ambiguities of video screens, Sluis developed a research collaboration with CSNI and expanded her activities to include the Unthinking Photography platform and a public programme of events in parallel with the 'exhibition' programme of the Media Wall, which she curated in collaboration with Sam Mercer. The Digital Programme of TPG adopted a networked research and collaboration approach, which increased its

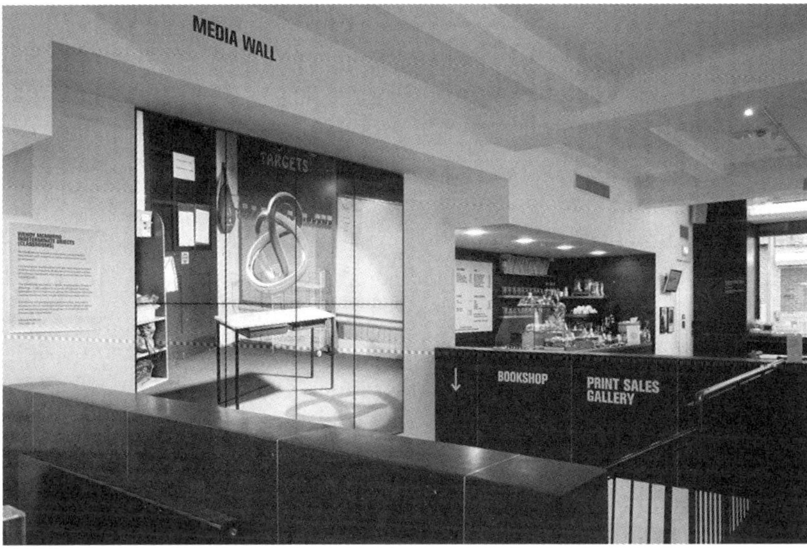

Fig. 8.1

Installation view of the Photographers' Gallery's Media Wall, featuring *Indeterminate Objects (Classrooms)* by Wendy McMurdo. 2017. Image © Kate Elliot.

reach and impact beyond the screen, supporting a collaborative PhD in which Nicolas Malevé undertook research on 'Algorithms of Vision'. Her ambition to take the culture of the Internet as a serious curatorial proposition needed a form of practice understandable by a photographic gallery, rather than as research, and to be located as part of an extended education and public events programme. To be curatorial a Digital Programme needed to be exhibitionary and the video screens in the revamped foyer presented the institution with a ready means of linking the screens to the digital image. In effect curating exhibitions for the screens was the entry point for the culture of the Internet and Sluis's increasing focus upon the computational image. The curatorial strategy of the Digital Programme used the video wall as the trojan horse for a wider range of initiatives designed to open up the politics of the network image.

The Media Wall comprised six high-definition 'Sharp' LED screens, mounted in a grid with an overall measurement of 3 x 2.8 metres, occupying a prominent space in the gallery entrance and café and highly visible from the street. Like many of the department and fashions shops in the surrounding West End of London, the video wall functions as a advertising device, which is both its strength and weakness when repurposed in the exhibitionary complex. Advertising screens call attention to something other than themselves, to the products they promote, which exist elsewhere, fleetingly present across the screen's parade of programmed images. Video walls can be ignored, passed by; they are incidental, part of the architectural background of the city. They are rarely an object of contemplation unless taken into the dedicated space of a gallery or inspected in a laboratory. The decision to site the wall in the entrance foyer was a choice of different exposures, that between an emanating beacon visible to all passing trade day and night, or a hidden gem sparkling in the inner sanctums of the galleries. In the commercial logic of media, the wall's position was a foregone conclusion. The wall would advertise TPG and its digital agenda. The distinction and hierarchy of space, a version of the above- and below-stairs of the aristocratic house, was contained in the director's interview with *The Guardian*.

> 'People still need a quiet space to look deeply at photographs and to reflect on their form and content, but there is also this tsunami of images on the internet and we, as a contemporary gallery, have a role to play in somehow making sense of that.' The Digital Wall, says Rogers, 'will reflect the new ways of curating, editing and re-imaging' that the internet has spawned, and 'will involve the public as co-producers of some of the work'. (O'Hagan 2011)

Since the first exhibition, *Born in 1987: The Animated GIF* (19 May–10 July 2012), curated by Sluis, a total of 36 exhibitions[25] have taken place. Of course, the Media Wall could have been programmed for any number of 'exhibits' over the same time period, but institutionally it conformed to the conventions of commissioning work, with a technical installation, opening event and a gallery schedule of four to six weeks. Within this convention the emphasis falls upon what types of project were shown.

The Media Wall exhibition programme has primarily tackled questions of the digital image in network culture by focusing upon the operations of software. In stark contrast the main exhibition programme of TPG continues to exhibit photography in its twentieth-century mode. Essentially the Digital Programme and its commissions fitted a new media agenda, drawing upon work in the new media art field. The work reflects a

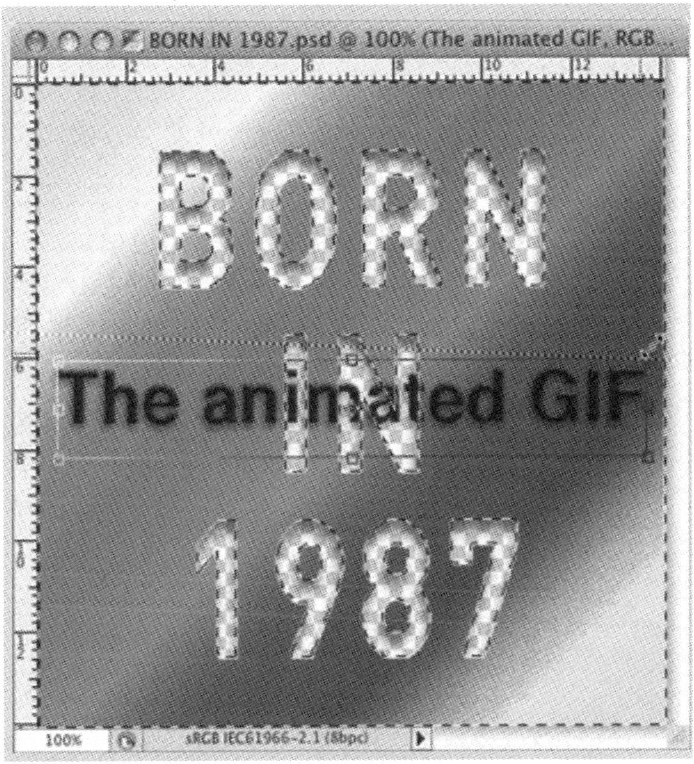

Fig. 8.2

Promotional image for *Born in 1987: The Animated GIF*. The Photographers' Gallery, London. 2012. Courtesy the Photographers' Gallery.

number of conceptual art strategies, including looking at the unintended consequences and glitches of software, looking at applications of surveillance technologies and social media and image sharing apps, the image as an embedded part of a platform and the image as hybrid of graphical animation. The Media Wall draws attention to the new conditions of the image online and on screens, looking at the flow of images and the software that produces them.

Katrina Sluis remains an adjunct research curator at TPG but in 2019 took up a full-time position at the School of Art at the Australian National University, leaving a sustained legacy, which awaits historical evaluation. Jon Uriarte, the second curator of digital programmes, has been building upon the legacy of the programme and initiating new strands. The programme currently has three strands: 'Imagin(in)g Networks,' looking at networks that use images to enable human and machine interactions; 'Data/Set/Match,' inaugurated by Sluis and representing an important shift away from the Media Wall towards a longer-term theme of investigation and activity and seeking new ways to present, visualise and interrogate image datasets; and 'Screen Walks,' a collaborative project with Marco de Mutiis, digital curator at Fotomuseum Winterthur, presenting a live-streamed artist/researcher-led exploration of online space and artistic strategies of digital cultures.

9

The Hybrid Image

Our intellectual life is out of kilter. Epistemology, the social sciences, the sciences of texts – all have their privileged vantage point, provided that they remain separate. If the creatures we are pursuing cross all three spaces, we are no longer understood. Offer the established disciplines some fine socio-technological network, some lovely translations, and the first group will extract our concepts and pull out all the roots that might connect them to society or to rhetoric; the second group will erase the social and political dimensions, and purify our network of any object; the third group, finally, will retain our discourse and rhetoric but purge our work of any undue adherence to reality – horresco referens – or to power plays.

Bruno Latour. 1993. *We Have Never Been Modern*. Cambridge, MA: Harvard University Press, p. 5.

This last chapter brings together the threads of the argument for forgetting photography. It considers how forgetting and remembering photography can become a productive programme to synthesise the various enquiries into the new conditions of the image and its politics. From the outset forgetting photography was proposed as a productive strategy for understanding photography afresh, by putting it to one side in order to see what it occludes of the current image condition. By adopting the perspective of photography's afterlife, or the zombie condition of photography, it becomes possible to locate the period to which it belonged and the world it showed, in a light no longer continuous with the present and revealing photography's once radical modernist promise now as a conservative force. The disjuncture between photography and the present image conditions becomes productive of the potential for a new conceptualisation of the image as well as a politics of a future 'digital' republic.[1] It is important to underline that the radical disjuncture between the photograph and what we will call here, still provisionally, the network image is not merely an extension of the earlier, narrow technological argument that the digital dealt the death blow to analogue photography. Such a position is, as has been pointed out by various scholars arguing for the cultural continuity of photography (Martin Lister, Sarah Kember et al.), a technologically determinist trap. But it is a trap which contains a second entrapment, in which the agency of technology is then diminished in order that culture can be seen as the more shaping force. Arguments have become progressively more sophisticated as media and cultural studies have expanded into critical studies of techno-science, as well as with the influence of a renewed political focus upon the material and intellectual degradation of the life world wrought by the forces of global capitalism.

Forgetting photography is an emphatic argument that the network image has little or nothing in common with the photograph. However, as previously discussed, the network image is an expansive epistemological apparatus, a relational socio-technical assemblage, which both limits and creates possibilities for how and what can be thought and imagined within it. The cultural myth of the persistence of photography and the endless empty repeat of the photographic trope is itself an organisation of desire, a receptacle for subjectivity as well as a disciplining of agency. The image in computational capitalism is the new burden of representation, first identified by John Tagg (1988) and importantly updated in Jonathan Beller's (2018) work, which sees media operating as a substrate of computational capitalism's extractive and exploitative organisation of labour. In essence, for Beller, media is a form of capitalist production, not a reflection of it, with a deeper systematic and organising relationship to everyday life, such that everyone now labours in the image. The political question for Beller's analysis is how totalising such affects are and whether and how they can be resisted.

The productivity of remembering photography from the perspective of its afterlife lies in the opportunities it presents for rethinking photography's complicity with the world and the world it objectified. This was the focus of zombie photography (Chapter 2) and the Tate and V&A case studies (Chapters 5 and 6), which essentially critique the ways in which histories and theories of photography in academia and museums maintain a canon of photography in the belief that it is still a contemporary medium. Remembering photography from the position of its afterlife literally opens up new ways of thinking about the realities, experiences and world of nineteenth- and twentieth-century Western culture and its colonial and imperial domination. As Beller's work suggests, photography was founded upon a racialised taxonomy and centrally effaced the barbarism of slavery in the name of a natural White order of the world. Remembering photography opens up many more opportunities to rethink the history of photography in relationship to nineteenth- and twentieth-century industrial capitalism and the ways in which photography was enlisted to paper over the cracks in its contradictory edifice. Remembering is also productive of rethinking photography as having always been an extended techno-social apparatus which defined bodies in time and space. Such an understanding was formulated by Ariella Azoulay (2012) as a civil contract, in which a photograph is a secular agreement of citizens involving different agents, none of whom individually has control of meaning. Azoulay has rethought photography in terms of a micro analysis of citizenship and public life, specifically situated by the dispossession of the Palestinian people by the Israeli state. Forgetting photography has a dual purpose, one the one hand it is productive of new perspectives on histories and cultures of photography, and on the other it looks forward to what has replaced photography, linked by the struggle to clarify the current moment. Unthinking and rethinking photography are the twin strategies for identifying and promoting a new politics of the image, called forth by the very conditions it attempts to identify.

There are many fruitful avenues in which forgetting photography can be carried forward into the organisation of knowledge, educational institutions, cultural policies and practices. Without pre-empting what forms future enquiries might take and recognising the complexity of the socio-technical image assemblage, the remainder of this chapter and the conclusion to the book discuss a number of epistemological and methodological issues which arise from attempting to think beyond photography. More provisionally still, it gestures towards an open agenda for future research and action. This conclusion is a restatement of the problems and a call to approach them in ways which are productive of the central agenda of planetary care and social justice. Media and visual cultural studies contribute to the formation of knowledge and how shared realities are constituted. Starting with the proposition of the image as a socio-technical assemblage, a number of questions need asking and are used in what follows and ultimately to signal an agenda for action.

First, there is a need to ask in detail what is included in the image as an assemblage and what its boundaries are, which will clearly involve stepping across existing humanities knowledge disciplines to embrace science, political economy, governance and diplomacy. Allied to this is the question of how the image is to be studied, not only in terms of a transdisciplinary approach, but also in terms of methods of enquiry, both practical and analytical, methods which need to stay close to the practices of the object itself. What then arises is a further question of in what conditions such work can be carried out, what institutional and financial arrangements need to be in place, what skills are needed and which people need to work together to facilitate a new kind of knowledge production. A lot follows from the answers to these first questions as conditions are never perfect, crossing boundaries is never easy and it is always in the knowledge that research models and funding currently follow instrumentalised agendas. This means that as yet there is still the task of making a convincing argument for bringing the object into view and for a different way of studying it. And yet much work has already been done across a range of initiatives in the fields of art and media practices, as witnessed by the topics of conferences and the focus of academic research centres, and one of the first tasks lies in sharing and linking such work. Studying the socio-technical image assemblage is itself part of the assemblage, in which linkages and collaborations, entailing questions of how the network operates and how it relates to apparatuses and institutional arrangements, are entailed. For the cultural sector, one primary question to be addressed is how the networked image can be curated and made public. This is a very practical and organisational question of how the university, museum and art organisation understand and relate to the Internet and computation. Finally for this study, we return to the question as to how the networked, relational and performative agency of the image operates upon and within media platforms and what purposes representation still serves in the multimodal and transmedial forms of global media culture. Such questions are more than enough to make a start and the discussion that follows briefly considers what is further involved.

Dimensions of the Socio-Technical Image Assemblage

computational (electronic)	hybrid (medial)	embodied (biological)
algorithmic (mathematical)	networked (cultural)	relational (connected)
data (signalled)	semiotic (historical)	spatial (immanent)

Fig. 9.1

The dimensions of the socio-technical image assemblage. 2021. Author.

Objects of Interest: From Network to Hybrid Image

The term image is complex and its use across this book has been subject to slippage in its frames of reference. It has required a number of highly specific qualifications, such that the reader might still remain unclear as to what is being defined and what falls within the parameters of the image as a socio-technical assemblage. It has to be insisted at this stage that the image after photography is a work in progress and that labelling runs the risk of reduction in a complex and expanding situation. Recall the terms reviewed in Chapter 6, in which the image was defined successively as digital, algorithmic, computational, operational, soft, networked and now hybrid. None of these terms taken on their own merits meet the threshold of the socio-technical image assemblage but taken together they build a model of its dimensions. The image is 'out there' in the world and 'in here' with us: a dynamic contingency of human vision; a received historical gathering of material objects embodying that which has been thought and seen; and a coupling of humans and machines. In the first sense the image shapes what it is possible to observe and to think; in the second sense images are material inscriptions of what has been observed; and in the third sense they are nonrepresentational forms of calculable data transmission. The image then can properly be called a hybrid which crosses and re-crosses nature and culture, art and science, human and non-human. The image in the computational mode of production is algorithmic, but it is also relational, embodied and semiotic. It is semiotic because of the paradox that the image is based upon a nonrepresentational socio-technical system but is made humanly understandable through representation. The representational and nonrepresentational are mutually dependent and cannot obliterate each other. Removing the residue of photography from representation opens up the image to a network of transactional and transmedial events (Thrift 2008).

Hybridity has been touched upon as a central concept for thinking the current complexities of the socio-technical image assemblage and serves as the chapter's title and motif in attempting to synthesis the many knowledge frameworks in which the image has been considered. In this respect Latour's analysis of the dichotomy of the moderns, between nature and culture, which created and sustains the endless epistemological and methodological war between purification and translation of hybrids, is highly instructive. Latour's model underpinned the understanding of the curation of photography at Tate and the collection of photography at the V&A in Chapters 5 and 6, for example.

In contrast to the profane world of the hybrid, in the history of art and cultural studies, the analogue work of art is considered a unique object, with the exception of fakes and copies, where fakes pass themselves off as unique and copies reinforce the uniqueness of the original. The guarantee of the purity of the original object is its singularity and its location within tradition. This was the analysis of Walter Benjamin (2008 [1939]) in saying that with the advent of photography and mechanical reproduction the aura of the original work of art withered because its image was ripped from the fabric of tradition by its mechanical reproduceability. With digital reproduction, and importantly the re-inscription of tradition as commodity, the relationship between art and photography has changed, paradoxically conferring singularity upon the analogue photograph and reducing further its distance from art. In the context of the museum we can see that Benjamin's analysis can be revised to say that in the era of computational reproduction, the photograph acquires a commodified aura, now as heritage, precisely because it is no longer at the centre of reproduction. Reproduction based upon a 'digital image' takes place through a distributed network of electronic switching, programmed by mathematical code. Paradoxically, a digital file continues to be culturally received as a singular object when rendered by a digital-to-analogue technical exchange as a printed image or data projected in an exhibition installation. Representational analysis is much harder to perform on the networked image, which is not only digitally composited by software, stored on relays of microchips, but also activated by its circulation and mobile set of coordinates to appear on a screen device. Initially this can be expressed by saying that the networked image is simultaneously a mathematical code rendered as a graphical animation in a digital video format, from which semiotic meaning continues to be read, and an encoded electrical transmission, relayed, received and read by a machine. The networked image is the result of a collaboration between human and machine. The appearance of an image in the network is dependent upon a series of nonlinear operations of coding, signaling and routing, together with software, platforms and viewing devices. It is in this sense that the networked image can be described as relational and performative. However, this initial description of the hybrid image prioritises the technical human/machine relations of the network of image formation and has not yet considered the first two senses in which the image is dynamic, immanent and historical. Hybridity defines the

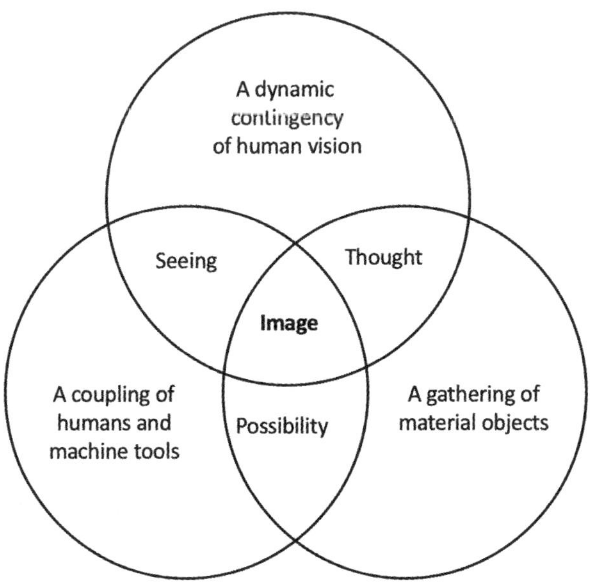

Fig. 9.2

The intersecting fields of the networked image. 2021. Author.

characteristics of networks, allowing 'cultural-technical' mingling in the human-machinic interaction. Hybridity allows us to grasp the image as the embodied practices of an image ecology. Conceptually, hybridity, coupled with networks, opens up meaning beyond singularity and representation into a multitude of subjective, social and political image practices and associations.

A hybrid is a mixture, an 'impure' intermediary category, combining different elements. In the case of the networked image the mixture is technical, social, temporal and spatial. The image thought of as a hybrid is a temporary object that a network assembles and makes visible. As Latour has argued hybrids are mixtures of nature and culture, which proliferate in the face of the constitutional work of humans in purification, which was earlier described as the process of photography's admittance to the museum. Latour also refers to hybrids as 'quasi-objects' (1993 p. 55) as a way out of the dualism between science and technology (nature) and society and the subject (culture). Quasi-objects are more social, fabricated and collective than the scientific view of natural objects, but equally they are more real, objective and non-human than the social sciences project society itself. As such the hybrid is defined by a heterogeneous mixture of nature and culture, operating in networks with no single point of origin and characterised by multiplicity. The hybrid image can therefore be thought of as present in many culture-nature 'entanglements' in which its agency can be identified as taking part in dynamic and embodied life worlds. If such a view were to be accepted as the condition of the image in culture it would require images to be discussed not in terms of singular representations or the mediation of given externalities, but as co-mingled subject/object hybrids. If the hybrid image reveals that mediation is itself the condition of the life world (Kember and Zylinska 2012), in which hybrids are manifest, then the image no longer stands outside of what it previously sought to represent, or mediate.

From the vantage point of the hybrid image, a number of things flow for a future research agenda. First, in revisiting photography, it becomes evident that historically the photographic image was formed by means of purification, as a taxonomic device, 'a cut' for ensuring that objects were sorted according to the constitution of nature and culture. Second, and in marked contrast, the networked image is constituted through the proliferation of hybrids of both nature and culture. The essential hybridity of the networked image, its 'quasi-object' status, leads to a conception of the image which accords with Braidotti's (2017) post-human view of the self, expressing multiple ecologies of belonging, while enacting transformation of sensorial and perceptual coordinates. The post-human perspective thus becomes productive of apprehending the image in the reflexive processes of consciousness, placing the image within experience and social practice. Such a definition also opens out onto a global transnational vista of cultural transformation brought into relief by the hybridity of the socio-technical image assemblage. For Arjun Appadurai (1996), for example, migration is a positive force in and of the network, operating against how the need for security is manipulated as a coercive form of control. Hence, the hybrid image realised as social movement, interaction and thought poses questions of control as well as agency in hypermodernity and the paradoxical present, as discussed in Chapter 8. Here we are thrown back onto the assemblage as both restricting the sensorial and perceptual co-ordinates of seeing and its standardisation and abstraction in the techniques of observation (Crary 1990) as well as the possibilities for global democratic extension of the network.

Hybridity suggests that understandings of images are relational and at present some of the connections, or lines of flight, in the network have not yet been traced. This is partly because such relations fall between the disciplinary interests of computing, in which the computer vision community operationalises data, as in the case of training machines to see, and media scholarship which lies outside of technological innovation, and analyses the social consequences. In both computer science and media analysis what is taken as seeing and reality remains based upon a representational image. In the case of computing the input and 'ground truth' is a photographic image database, and in media the output is studied as the continuing social practices of photography. The black box of the technical apparatus and the extensive invisible networks of circulation standing between these two sets of practices are only dimly coming into view. Few computer scientists see the relevance of cultural questions and few culturalists understand computer coding. What is argued for here, in order to finally forget photography as the contemporary register of the image and move on, is to reorganise and mobilise knowledge in such a way that it can grasp this new situation fully, to be able to follow the ways in which value is produced by the cooperation between humans and machines in the network.

Knowledge Paradigms and the Academy

There is something ultimately arbitrary in the selectiveness of theoretical reference and there are many ways of framing the problem of rethinking and unthinking photography. Latour provides a convincing account of the paradoxes of modern thought, which can be applied to and illuminate the predicament of the image beyond photography. In his critique of positivist and progressive sociology and the development of actor network theory, he provides a method for rethinking what the image has become in terms of a network or assemblage.

In attempting to move beyond photography and open up new understandings of the relationship between representation, reproduction and the real, the networked image remains an elusive object, traversing a number of humanities sub-fields. In an important sense the networked image, as the most obvious manifestation of the socio-technical image assemblage, can be considered a boundary object, providing a liminal zone and interface between discourses and the objects framed by their field of view and the worlds they describe, but not belonging to any of them (Star and Griesemer 1989). The problem then becomes understanding what the limits as well as the possibilities are when considering the networked image within different knowledge fields. Looking at how photography has been inscribed within the interests of art history, philosophy and media has shown how each field excludes meaning in order to bound the object and in the process abstracts photography from the life world at the risk of reification. The institutionalisation of photography has also shown how collecting and exhibiting photography decontextualises the object through the purification of the image. Such institutional and epistemological procedures succeeded in fracturing the object of photography, parsing selective understandings and turning away from public education. In contrast, opening up the discussion of the world after photography presents a new opportunity to consider the image in public, with all of the problems associated with it, its truth, veracity, construction and consumption in the light of hybridity. As discussed in the two previous chapters, serious attempts are being made in the arts and humanities to grasp the conditions of the networked image, but it is also possible to see that the networked image, like the photograph, can be divided up into different disciplinary frames. There is everything to be gained from different perspectives and competing arguments, as long as they are debated in the same room, but there is the very real danger and much evidence that that is not happening under the conditions of the commodification and privatisation of knowledge, which this project has laid out from the beginning. The commodification of knowledge eviscerates the common room and public space and the network is also an agent of commodification. To combat such instrumentalisation of knowledge, in which the networked image is treated as information rather than as politics, a new approach to research and knowledge formation is needed, even if that is only possible in the interstices of institutions and against the grain of the new silos of the Internet. It must also be recognised that local and international attempts to assemble to share understandings have been and continue to be made in many gatherings, Transmediale[2] being only one such positive example. But from the argument made here the framework of discussion needs widening and new disciplines, actors and actants need to be admitted. In order to advance this cause future research needs to be organised and constituted differently and to do that hard and persuasive arguments will have to be made to politicians, policymakers, research councils and vice chancellors of universities and ultimately to Google and Microsoft. For arts and humanities scholarship to stop trailing technology, dealing only with its after-effects, it needs to adopt a model of social science which no longer separates nature and culture. In this way arts and media practices could be part, if not at the forefront, of technological research and development.[3] But to get anywhere near this objective, the arguments for why we need to look at the hybrids of the network needs to be heard across all of the communities of interest. If such a reorganisation was achieved, then both software development and its media uses would be driven by a different order of priority and a new accountability would emerge. Such a project has something of Latour's idea of a parliament of things, based upon a non-modern constitution whose first two guarantees are 'the non-separability of the common production of societies and natures and the continuous following of the production of Nature, which is objective, and the production of Society, which is free' (1993, p.142). As Latour argues,

> what does it matter, so long as they are all talking about the same thing, about a quasi-object they have all created, the object-discourse-nature-society whose new properties astound us all and whose network extends from my refrigerator to

the Antarctic by way of chemistry, law, the State, the economy, and satellites. The imbroglios and networks that had no place now have the whole place to themselves. They are the ones that have to be represented; it is around them that the Parliament of Things gathers henceforth. (1993, p. 145)

But, as the book has argued, academia plays an active part in the cultural conundrum of the continuing afterlife of photography and is deeply embroiled in its reproduction. This it has been argued is a consequence of the divisive, competitive and myopic conditions of current knowledge production, which ironically are for the most part unreflexively bracketed out of scholarly work. This is why a new approach is needed in which knowledge can become common and public by embedding research in independent collaborative communities of practice beyond the university. A visible paradox within cultural, art and media studies is that while there is a fashionable churn of theory, promising new vistas and radical agendas for doing things differently, the cultural and institutional form of knowledge production remains the same. Radical critique is reduced to career development and promotion, whilst the increased emphasis upon external collaboration is orientated to markets and commodities. Academic trends have been accelerating over the past decade, in a cycle of cultural over-production, creating greater theoretical redundancy. It is not that we can do without critical thought, but rather a recognition of the increasingly severe limits on the self-serving mode of the reproduction of critical theory. This is a pragmatic view and a post-critical position,[4] which can easily be mistaken for a rejection of theory altogether and in the extreme can be enlisted by populist anti-intellectualism. Such is the precarity of the world as well as the current academic plight, in which, under a global neoliberal regime, academic research has been instrumentalised to serve short-term market interests. For the humanities, such instrumental thinking has squeezed independent critical thought to the margins of funding, but has also in instrumental systems brought critique to its limit. But the alternative is just as problematic, in which the assessment of research for funding purposes in the UK, the six-yearly cycle of the research excellence framework (REF), is used to drive academics to increase research outputs. Both of these mechanisms are designed to make academic research accountable to its public funding and still based upon peer review, which is a laudable aim, but this is not the same thing as making research public in the sense of common ownership, which is a better and more important aim.

For the position attempted in this book one complex example of cultural churn and theoretical fashion, seen in current postgraduate and doctoral work, can be found in theoretical discourse which advocates non-binary theoretical approaches, encouraging perspectives which foreground relational, complex entanglements in the world, rather than the Enlightenment rationalism of subject and object. The current epistemic tension is a legacy of postmodernism, the continuation of the crisis of representation and of the disarray of Left politics. One such attempt at a non-binary method of seeing the world can be seen in the work of Karen Michelle Barad (2007), in which she puts forward the concept and theory of agential realism. Barad's analysis, drawn from the quantum physics of Niels Bohr, offers a radically different way of understanding the universe as phenomena made through the intra-acting agencies of human and non-human matter. For Barad, human agency is not a given property of the mind and body, but emerges as a condition of possibility of phenomena. Barad's work has inspired many scholars and contributes to a growing global movement of post-anthropocene politics, which demand that humanity prioritises care for its planetary home, but needs to think differently about how it understands phenomena. The same can be said for the success of science and technology studies (STS) in showing the complex knowledge entanglements between humans, their material technical systems and the biosphere (Bruno Latour, Michel Callon, Isabelle Stengers, and John Law), which also runs counter to the entrenched specialisations of disciplinary expertise in politics, economics, science and the arts and the separations within them of theory, practice and policy. Meanwhile the institution in which such theories are developed and intellectuals given a home since the Enlightenment moves slowly on, impervious, it appears, to the illuminations of theory. The concrete life of the university remains stubbornly organised according to taxonomies and disciplines that maintain hard separations between nature and culture and the life of the mind and that of the institution. Theory does not translate directly into reflexive modes in either personal or institutional life.

This discussion may seem an overly grand and distant perspective to conclude a book on the place of photography in contemporary visual culture and of course, on the established view of knowledge and its taxonomies, photography would rank very low on any scale of planetary urgency. But that's the point – as long as academia and education more generally continue to approach the world through ever more hierarchical sub-divisions of instrumentalised subject knowledge, the more the paradox of an excessively knowing world, knowing it needs to change, but collectively not knowing how, is replicated. Bernard Stiegler (2016) gives a greater urgency to this state of affairs in saying that the survival of universities depends upon them taking over the digital transformation if they wish to be promulgators of knowledge, rather than mere dispensers of education. If the university doesn't become digital is will no longer exist, argues Stiegler, but to do this the university must exercise its socio-political and economic responsibilities, and create an alternative publishing industry to that of Google and Amazon. Without such a movement within institutions the humanities will be relegated at best to conservative cultural heritage and at worst to an irrelevance.

The Curation of the Network Image, Public Culture and Ways of Seeing

Across Europe, the aggressive form of neoliberal capitalism has dismantled and continues to dismantle and deregulate the civic society of the reforming late-nineteenth-century and twentieth-century social democratic welfare state. Such attacks on democracy take the overt form of the privatisation of the public realm and the normalisation of the hyper-individual, both carried out under the cover of economic and scientific progress. For culture and education, the consequences of neoliberalism are to be found in commodification and competition. Over the same period, photographic practice, for the most part, has marched to the same neoliberal tune, as discussed in Chapter 8, evidenced in the spectacular enfolding of photography into the global art market on the one hand and its further enlistment into the commodification of the self on the other. In hypermodernity photography repeats itself, probably as farce, as the modern medium of the industrial revolution. Modernity contains both the notion of the ascendency of temporal progress, time's irreversible arrow, and a revolution in time in the cultural rupture with a stable archaic past, which Latour associates with the modern preoccupation with mastery, domination and emancipation. Over the course of the twentieth century photography took part in modernism's masterful view of social and scientific progress based in technology and social democracy. However, in order to be a medium of the modern, in Latour's terms, photography had to engage in the work of purification, not only classifying the world according to the distinct ontological zones of the modern, but also to provide its own ontological purification as a modern medium. Such a state of affairs explains the invention of genres of photography, as an attempt at medium specificity, in the face of the many hybrids of photography proliferated by reproduction. Today the work of purification continues apace across the institutions of art, academia and museums, and yet the more photography is purified, the more the hybrids of the networked image proliferate. The dichotomy thus leads to a contradictory state of affairs in which the modern requires the critical project of purification, even though that project is developed only through the proliferation of hybrids. Latour argues that modernity is thrown into question by recognising both hybrid networks and the work of purification and that 'the more we forbid ourselves to conceive of hybrids the more possible their interbreeding becomes'.

In the afterlife of photography, photography is everywhere but no longer socially relevant, because it is consumed in a commodified form and without a public space. This is the political situation faced in any attempt to curate, archive and exhibit the network image as photography's substitute. Attempts to rescue photography as capable of acting back upon its alienating consequences as unpaid labour were the subject of considering Ben Burbridge's conception of the possibility of a post-capitalist photography. Burbridge sees the possibility of photography operating as a means of sharing in a post-work world and opening up the possibilities for building a post-capitalist world in the present. Burbridge understands the changed technological circumstance in which visual images are made yet retains the myth of photography as a popular cultural currency based upon computational practices.

Attempts to move beyond photography were the aim of Katrina Sluis's Digital Programme at the Photographers' Gallery (TPG) between 2012 and 2019, in which she shifted the discussion and curatorial practice decisively to the algorithmic and computational image. Sluis's model of curation at TPG had a base in academic research and as such had theoretical as well as practical reach, which gave her the vantage point of understanding the limits of photographic culture and the problems of moving beyond it. Her paper, with Daniel Rubinstein (2013a), on the algorithmic image originated, as she says, 'in a frustration which Daniel Rubinstein and I had with the persistence of certain approaches to the photographic image which fail to account for the temporality and excess of a photographic culture which has become algorithmic or "softwareised"'. But as she recognised as a curator, such an analysis bequeaths the problem of how software can become the visible subject of curation, or, as she puts it,

> How do we engage the public with the politics of metadata without resorting to data visualisation? How do we engage with the dynamism of the computational camera without turning it into a dead object? Ultimately I'm not sure the exhibition is always the right site through which these dynamics can be addressed. (Zorzal and Menotti 2017)

Both Burbridge and Sluis were seeking a public platform for an image politics capable of both revealing and critiquing the socio-technical image assemblage's new forms of subjectification as well as seeking within it progressive alternates. Such alternatives were most readily found in new media art practices critically investigating the social, economic and military applications of imaging technologies, rather than engaging with the social and cultural practices of network users. The reliance on the art field as an exclusive site of criticality betokens a continued problem of how to recognise and engage new publics in old public institutions. Public institutions of exhibition have had to adapt to privatisation, which paradoxically orientates them to the 'creative force' of the market, with the result that 'modern' critical public culture becomes commodified and consumed as a heritage product. This was the analysis presented in Chapter 5 in discussing photography at Tate Modern and Tate Britain. Here the conclusion, in the same manner as Stiegler's call for the university to 'be digital', was that public museums and contemporary public galleries also need to embrace the network as their modus operandi, rather than its current organisational annexation to marketing, or exhibited as a form of art that uses media. The zombie state of photography perpetuates modernity's temporal linearity in the fiction of the contemporary and the presentation of media as discrete tools for expressive use. Such views lead to a profound misrecognition of the socio-technical image assemblage as a cultural value system, as a hybrid waiting to be admitted rather than purified.

Media and the Socio-Technical Image Assemblage

The book has focused exclusively upon the situation facing photography and the photographic image in its consideration of the new mode of image production and has not attended to the parallel remediations, outlined by Bolter and Gruisin (2000) of film, cinema, print, video and television, which arguably have much greater force in the global media sphere. However, it is possible to apply the account of the radical reconstitution of the image to other previously analogue mediums, most obviously film. Other studies have focused upon the same socio-technical image assemblage from the perspective of other media (Cubbitt 2021) which in one way or another attempt to outline the defining features of a new medium. Paul Frosh (2019) argues that digital media manifests the given reality through encounters and disclosures. Daniel Strutt's (2019) work on post-cinema argues that when new visual technologies arrive to represent and simulate reality, they give rise to nothing less than a radically different sensual image of the world. As noted by Marie and Hoelzl (2014), the JPEG image, so closely associated with the photograph, is in fact a video format. This volume has already traced the interrelationship between photographic and new media discourse in Chapter 3 in the frame of post-photography, where an attempt to forge a field of study, based on digital convergence, has been continually outstripped by the speed of events. Media studies has worked to reconstitute the objects of media as a response to the radical changes taking place under the sign of computational capitalism and automation, but visual cultural studies has an older claim to deal with what Crary earlier defined as the 'the plural forces and rules composing the field in which perception occurs'. Nicholas Mirzoeff (1999) made a compelling case for establishing a new field of the study of visual culture, arguing that the increasing importance of the visual in culture required a different optic from that of separate specialist studies of images. Mirzoeff argued that visuality had a new dominant and organising political function in contemporary society and proposed to view modern media, art, photography, film and television collectively in order to identify the underlying logic and programme of visualisation. For a time, visual cultural studies appeared as if it might claim the space for a new unifying field of knowledge, where discrete studies of visual media might be brought together and expanded in to a general theory, in which the social, historical and technical optics of visuality and its apparatuses could be understood. Visual culture gained in popularity over a period up to and including the present. It gained institutional legitimation, expanding into academic departments, research groups and named courses in the same manner that British cultural studies had developed from the 1960s. Over time and with the increasing commodification and instrumentalisation of knowledge production, the conjunctural political project of visual culture, like that of cultural studies, dissipated. In place of an explicit politics, the idea of visual culture came under critical academic scrutiny, seen to lack a common paradigm and remain an ambiguous term. As Karin Becker (2004 summed it up, 'The term itself is ambiguous, implying on the one hand that culture as we know it is in the throes of becoming more visual, requiring thereby new theories and modes of inquiry to under-stand it. On the other hand, the term suggests that the new field is directed toward visual culture and its artefacts, isolating them from other cultural forms' (2004, p. 150).

Central as critical scrutiny remains, visual culture as an institutional intervention has become subject to academic turf wars over which objects belonged in which domain and this seems to fit with the critique offered throughout this book, as being a symptom of the overproduction of knowledge, which keeps the wheels of academia turning.

As for Nicholas Mirzoeff, who now describes himself on YouTube as a visual activist,[5] the cultural politics of the term visual culture gained greater purchase outside of academia. Mirzoeff (2011) expanded and conceptualised his original view of the project of visual culture in *The Right to Look: A Counterhistory of Visuality*. In his expanded schema, Mirzoeff, like Crary, Frosh and Appadurai, defines visualisation as an imaginary, because as he says the 'practice must be imaginary, rather than perceptual, because what is being visualized is too substantial for any one person to see and is created from information, images, and ideas' (2011, p. 474). Here the imaginary is also abstracted to the level of a system, but one located in materialist history and struggles for emancipation. Mirzoeff defines 'complexes of visuality', which he sees as corresponding to historical development in a similar manner in which Jonathan Beller earlier analysed photography in relationship to slavery. Mirzoeff defines the functions of the complexes of visuality as classifying, separating and aestheticising and he identifies three separate successions, starting with the plantation complex from 1660 to 1865, followed by the imperial complex from 1857 to 1947 and finally the military–industrial complex from 1945 to the present. Mirzoeff's schema works in the manner of Michel Foucault's genealogies, tracing and revealing the formations of power/knowledge, drawing attention to the ways in which visualisation operates in reproducing the social relations of the mode of production in which domination and oppression still prevail. Importantly Mirzoeff articulates the possibilities of counter-histories and thus arrives at the right to look. Mirzoeff's stance as a visual activist is a continuation of the public intellectual, who understands knowledge as political and therefore has an ethical duty as well as a right to speak. John Berger came from the same tradition and Jonathan Beller's work follows the same path of staying with capitalism's deployment of media as ideological apparatuses, only given much greater sophistication as computational capitalism, which invades, shapes positions subjectivities ever more extensively.

Computation as predicted by Jonathan Crary (1990) has 'reconfigured the relationships between an observing subject and modes of representation that effectively nullifies most of the culturally established meanings of the terms *observer* and *representation*.' What has changed, argues Crary, 'are the plural forces and rules composing the field in which perception occurs' and in which vision is determined by 'the functioning of a collective assemblage of disparate parts on a single social surface'. Going even further, Crary thinks, 'It may even be necessary to consider the observer as a distribution of events located in many different places'. Over the three decades that have elapsed since Crary's *Techniques of the Observer*, his incisive and meticulous analysis has proved remarkably accurate. If the observer is now understood as a distribution of events differentially located, then so too is the image as it is met in the socio-technical assemblage. Hito Steyerl (2013) argues that humans are now embedded in images and in her essay 'Too Much World: Is the Internet Dead?', published in *e-flux*,[6] she sees the image as having crossed the screen and into reality. Images are no longer representations of a pre-existing condition, but rather, 'Data, sounds, and images are now routinely transitioning beyond screens into a different state of matter. They surpass the boundaries of data channels and manifest materially. They incarnate as riots or products, as lens flares, high-rises, or pixelated tanks. Images become unplugged and unhinged and start crowding off-screen space. They invade cities, transforming spaces into sites, and reality into reality.'

In a further essay, 'In Defence of the Poor Image' (2009), Steyerl makes the point that it is a mystification to think of the digital image as an immaterial clone of itself, insisting on the degraded material inscription of its copy, bearing the traces of its rips and transfers, a copy in motion, as she puts it, one which constructs anonymous global networks, building alliances as it travels and creating new publics and debates.[7] The condition of the hybrid image illustrates that the question of representation and meaning is henceforth transposed to networks of circulation and the mobility of subjects in which encountering images is deregulated, arbitrary, excessive, polysemic and transmedial. How in this networked situation are the new alliances to be made sense of and how do they relate to what remains of the ethics and values of the public media and its spaces of viewing?

One of the preliminary tasks in Bruno Latour's parliament of things would be to connect digital media studies scholarship, which continues with a representational analysis of digital media in terms of social shifts in use and behaviour, and cultural studies, which takes everyday material life as its object, with the various groups of scholars in software studies, cybernetic studies and network studies who are on the trail of the algorithm in seeking the meaning system of nonrepresentation. In summary, in order to make the socio-technical image assemblage visible, research and scholarship needs to redraw the boundary of the object, leading to a transdisciplinary and transmedial approach. This task needs to be carried out across theory, practice and policy in collaboration with progressive collectives who are already attempting to recognise the hybrids in working to ensure a democratic and sustainable future. Purification has led to risk avoidance and the paradoxical practices of conservative modernising, whereas the hybrid requires risk to be embraced in order for the future to be secured. In making this argument, the stance taken of photography forgotten led in turn to a perspective of its recovery. At the outset it was stated that forgetting photography is not about one medium superseding another, nor the remediation of the medium of photography by computing, although that is certainly happening. Both zombie photography and computational imaging exist in the same temporal space and the effort is to understand the conditions that maintain their state and the relations between them.

And Finally

The epistemological thread of forgetting photography started, as it has for much other cultural analysis, with Foucault's critique of positivist and idealist intellectual absolutes and certainties about the world, the individual, society and the nature of power. But Foucault was referenced at the beginning of this account only as its point of departure, through Jean Baudrillard's essay entitled *Forget Foucault,* originally published in 1977. In a much later essay, 'Requiem for the Twin Towers,' Baudrillard (2012) was not insisting, as Latour attacked him for, that the material world is of no consequence, vanished as it were, although he is pointing to reality as an image. Rather, he was demonstrating his method of fatal strategies, which parallels Latour in questioning the limits of critique, but with radically different intellectual outcomes. Whereas critical theory's task is to master the object, the fatal theorist seeks to anticipate the irony of the object. As Kip Kline (2016) points out, in Baudrillard's simulated world everything is actualised and transparent with no more illusion. The foundation of fatal strategies can be seen in the 'Forget Foucault' essay (1977, which was submitted to and rejected by the highly esteemed French academic journal *Critique,* of which Foucault was an editor. The essay is structured around Baudrillard's motif of the murder of the real, but its real object was the established theoretical architecture of the real, represented by Foucault's work. 'Forget Foucault' opens with Baudrillard declaring Foucault's writing to be too 'perfect' in giving an account of what it proposes. Baudrillard argues that as a discourse of power, no longer based on a despotic or catastrophic architecture, Foucault's writing is a seamless, meticulous unfolding of a narrative without origin, in which power 'seeps through the whole porous networks of the social, the mental and of bodies infinitesimally modulating the technologies of power' (2007, p. 30).

In short, Baudrillard sees Foucault's discourse as a mirror of the powers it describes which cannot be taken as a discourse of truth, but a mythic discourse, which has no illusions about the effect of the truth it produces. For Baudrillard, Foucault's writing is 'too beautiful to be true' and its very exactitude and flawlessness in understanding discipline, power, sexuality and the body maintains an unquestioning objective reality principle at a point in time when power escapes us, 'made hyperreal through simulation' (2007, p. 31).

The polemic advanced over the course of this book has been that photography in its analogic perfection, like Foucault's writing, has been 'too perfect' and, in the same manner as Foucault's discourse of power, zombie photography continues to reinforce an historical objectivity based upon ensuring the reality principle of capitalism. However, no sooner has zombie photography been critically dispatched than another head of the Hydra appears, this time in the form of the network image in computational culture and its alienating algorithmic apparatus. The first strategic task for critical thought and progressive politics is to understand what reality is conjured by this new apparatus and the second task is to work out a set of tactics for engaging with and finding productive entries into the larger socio-technical image assemblage. This task will be made immeasurably easier by adopting an approach which remembers, reconciles and finally forgives photography.

Bibliography

Adams, Ruth. 2010. 'The V&A: Empire to Multiculturalism?', *Museum and Society*, 8(2), pp. 63–79.

Agamben, Giorgio. 2009. *"What is an Apparatus?" and Other Essays.* Stanford, CA: Stanford University Press.

Appadurai, Arjun. 1996. *Modernity at Large: Cultural Dimensions of Globalization.* Minneapolis, MN: University of Minnesota Press.

Azoulay, Ariella. 2012. *The Civil Contract of Photography*. Princeton, NJ: Zone Books.

Baker, George. 2005. 'Photography's Expanded Field', *October*, 114, pp. 120–140.

Bal, Mieke. 2003. 'Visual Essentialism', *Journal of Visual Culture,* 2(1), pp. 5–32.

Banerjee, S.B. 2008. 'Necrocapitalism'. *Organization Studies*, 29(12), pp. 1541–1563.

Barad, Karen Michelle. 2007. *Meeting the Universe Halfway: Quantum Physics and the Entanglement of Matter and Meaning*. Durham, NC: Duke University Press.

Bärnighausen, Julia, Caraffa, Costanza, Klamm, Stefanie, Schneider, Franka and Wodtke, Petra. 2019. *Photo-Objects: On the Materiality of Photographs and Photo Archives in the Humanities and Sciences.* Munich: Max Planck Society.

Barthes, Roland. 1977. 'The Rhetoric of the Image', in *Image, Music, Text,* ed. and trans. Stephen Heath, pp. 32–51. New York: Hill and Wang.

Barthes, Roland. 1984. *Camera Lucida*. London: Fontana.

Barthes, Roland. 1993. *A Barthes Reader*, ed. Susan Sontag. London: Vintage Classics.

Batchen, Geoffrey. 1992. 'On Post-Photography', *Afterimage*, 20(3), pp. 17–24.

Batchen, Geoffrey. 1999. *Burning with Desire*. Cambridge, MA: MIT Press.

Batchen, Geoffrey. 2000. *Each Wild Idea*. Cambridge, MA: MIT Press.

Batchen, Geoffrey. 2002. 'Proem', *Afterimage*, 29(6), p. 3.

Batchen, Geoffrey. 2004. *Forget Me Not: Photography and Remembrance*. New York: Princeton Architectural Press.

Batchen, Geoffrey. 2021. 'Photography in the Age of Massification', correspondence between Joan Fontcuberta and Geoffrey Bartchen, in *Photography Off the Scale*, ed. T. Dvorak and J. Parrika, pp. 253–287. Edinburgh: Edinburgh University Press.

Bate, David. 2010. 'The Memory of Photography,' *photographies,* 3(2), pp. 243–257.

Baudrillard, Jean. 1981. *For a Critique of the Political Economy of the Sign*. New York: Telos.

Baudrillard, Jean. 1983. 'The Ecstasy of Communication', *in Postmodern Culture*, ed. Hal Foster pp. 126–134. London: Pluto.

Baudrillard, Jean. 1994 [1983]. *Simulacra and Simulation*, trans. Sheila Glaser. Ann Arbor, MI: University of Michigan Press.

Baudrillard, Jean. 2000. '*Photography, Or The Writing Of Light*', trans. François Debrix. https://journals.uvic.ca/index.php/ctheory/article/view/14605/5462.

Baudrillard, Jean. 2005. *The Intelligence of Evil or the Lucidity Pact*, trans. Chris Turner. Oxford: Berg.

Baudrillard, Jean. 2007. *Forget Foucault*, trans. Nicole Dufrense. Cambridge, MA: Semiotext(e).

Baudrillard, Jean. 2012. *The Spirit of Terrorism and other Essays*. London: Verso.

Beck, Ulrich. 1992. *Risk Society: Towards a New Modernity*. London: Sage.

Beck, Ulrich. 2002. *Power in the Global Age: A New Global Political Economy*. Cambridge: Polity.

Beck, Ulrich. 2003. 'Toward a New Critical Theory with a Cosmopolitan Intent,' *Constellations*, 10(4), pp. 453–468.

Beck, Ulrich and Beck-Gernsheim, Elisabeth. 2001. *Individualization: Institutionalised Individualism and its Social and Political Consequences*. London: Sage.

Beck, Ulrich, Giddens, Anthony and Lash, Scott. 2007. *Reflexive Modernization: Politics, Tradition and Aesthetics in the Modern Social Order*. Cambridge: Polity.

Becker, Karin. 2004. 'Where is Visual Culture in Contemporary Theories of Media and Communication?,' *Nordicom Review*, 25(1–2), pp. 149–157.

Belden-Adams, Kris. 2019. *Photography, Temporality, and Modernity: Time Warped*. London: Routledge.

Bell, David and Kennedy, Barbara. M. (eds). 2000. *The Cybercultures Reader*. London: Routledge.

Beller, Jonathan. 2006. *The Cinematic Mode of Production: Attention Economy and the Society of the Spectacle*. Dartmouth, NE: Dartmouth College Press.

Beller, Jonathan. 2016. 'Informatic Labor in the Age of Computational Capital,' *Lateral: Journal of the Cultural Studies Association*, 5(1). https://csalateral.org/issue/5-1/informatic-labor-computational-capital-beller/.

Beller, Jonathan. 2018. *The Message is Murder: Substrates of Computational Capital*. London: Pluto Press.

Belting, Hans. 2014. *An Anthropology of Images: Picture, Medium, Body*. Princeton, NJ: Princeton University Press.

Benjamin, Walter. 1931. *A Short History of Photography*. https://monoskop.org/images/7/79/Benjamin_Walter_1931_1972_A_Short_History_of_Photography.pdf, trans. (1977) Patton, Phil. In *Artforum*, 15(6).

Benjamin, Walter. 2008 [1939]. *The Work of Art in the Age of Mechanical Reproduction.* London: Penguin.

Bennett, Tony. 1988. 'The Exhibitionary Complex', *New Formations* 4, Spring. http://banmarchive.org.uk/collections/newformations/04_73.pdf.

Bennett, Tony. 2013. *The Birth of the Museum.* London: Routledge.

Berger, John. 1972. *Ways of Seeing.* London: Penguin.

Berger, John. 2009. *About Looking.* London: Bloomsbury.

Berger, John. 2013. *Understanding a Photograph.* London: Penguin.

Berry, D. 2011. *The Philosophy of Software: Code and Mediation in the Digital Age.* Basingstoke: Palgrave Macmillan.

Bhabha, Homi K. 1994. *The Location of Culture.* London and New York: Routledge.

Bishop, Claire. 2012a. *Artificial Hells, Participatory Art and Politics Spectatorship.* London: Verso.

Bishop, Claire. 2012b. 'Digital Divide: Contemporary Art and New Media', *Artforum*, September. www.artforum.com/print/201207/digital-divide-contemporary-art-and-new-media-31944.

Bishop, Claire. 2013. *Radical Museology, or, What's Contemporary in Museums of Contemporary Art?* Cologne: Konig, Verlag der Buchhandlung.

Bolter, Jay David and Gruisin, Richard. 2000. *Remediation: Understanding New Media.* Cambridge, MA: MIT Press.

Bourdieu, Pierre. 1965 [1996]. *Photography: A Middlebrow Art.* Cambridge: Polity.

Bourdieu, Pierre. 1996. *The Political Ontology of Martin Heidegger.* Cambridge: Polity.

Bourdieu, Pierre. 2010. *Distinction: A Social Critique of the Judgement of Taste.* London: Routledge.

Boutang, Yann Moulier. 2012. *Cognitive Capitalism.* Cambridge: Polity.

Braidotti, Rosi. 2017. *The Posthuman.* Cambridge: Polity.

Bratton, Benjamin H. 2015. *The Stack: On Software and Sovereignty.* Cambridge, MA: MIT Press.

Brett, Donna West and Lusty, Natalya (eds). 2019. *Photography and Ontology: Unsettling Images.* London: Routledge.

Bridle, James. 2018. *New Dark Ages: Technology and the End of the Future.* London: Verso.

Brückle, Wolfgang and de Mutiis, Marco. 2019. 'Post-Photography: What's In a Name?'. *Spectrum: Photography in Switzerland*, September 19/December 2020. https://photography-in-switzerland.ch/essays/post-photography-whats-in-a-name.

Burbridge, Ben. 2020. *Photography After Capitalism.* London: Goldsmiths/MIT Press.

Burgin, Victor (ed.). 1982. *Thinking Photography*. London: Macmillan.

Bush, Vannevar. 1945. 'As We May Think', *The Atlantic*, July.

Caines, Matthew. 2013. 'Arts Head: Amit Sood, Director, Google Cultural Institute', *The Guardian*, 3 December. www.theguardian.com/culture-professionals-network/culture-professionals-blog/2013/dec/03/amit-sood-google-cultural-institute-art-project.

Callon, Michel. 1980. 'Struggles and Negotiations to Define What is Problematic and What is Not: The Socio-logics of Translation', in *The Social Process of Scientific Investigation*, ed. K. Knorr, R. Krohn and R. Whitley, pp. 197–219. Dordrecht: D. Reidel Publishing Co.

Castells, Manuel. 2000. *The Rise of the Network Society: Economy, Society and Culture v.1.* London: Wiley-Blackwell.

Chun, Wendy Hui Kyong. 2011. *Programmed Visions: Software and Memory*. Cambridge, MA: MIT Press.

Colebrook, Claire. 1999. *Ethics and Representation: from Kant to Post-Structuralism.* Edinburgh: Edinburgh University Press.

Colebrook, Claire. 2014. *Death of the PostHuman: Essays on Extinction, Vol. 1.* London: Open Humanities Press/Michigan University Press.

Crary, Jonathan. 1990. *Techniques of the Observer: On Vision and Modernity in the 19th Century*. Cambridge, MA: MIT Press.

Crouch, Colin. 2004. *Post-Democracy (Themes for the 21st Century)*. London. Polity.

Cubitt, Sean. 2014. *The Practice of Light: A Genealogy of Visual Technologies from Prints to Pixels*. Cambridge, MA: MIT Press.

Cubitt, Sean. 2021. 'Mass Image, Anthropocene Image, Image Comnmons', in *Photography Off the Scale*, ed. T. Dvorak, and J. Parikka, pp. 25–40. Edinburgh: Edinburgh University Press.

Cubitt, Sean, Palmer, Daniel and Tkacz, Nathaniel (eds). 2014. *Digital Light.* London: Open Humanities Press.

Cumming, Laura. 2014. 'Conflict, Time, Photography Review – Emptiness, Ruin and Absence', *The Guardian*, 30 November. www.theguardian.com/artanddesign/2014/nov/30/conflict-time-photography-review-tate-modern-emptiness-ruin-absence.

Dahlgren, Anna, Petersson, Dag and Vestberg, Nina Lager. 2013. *Representational Machines: Photography and the Production of Space.* Aarhus: Aarhus University Press.

Dean, Jodi. 2005. 'Communicative Capitalism: Circulation and the Foreclosure of Politics Cultural Politics', *Cultural Politics: An International Journal*, 1(1), pp. 51–74.

Dean, Jodi. 2009. *Democracy and Other Neoliberal Fantasies: Communicative Capitalism and Left Politics.* Durham, NC: Duke University Press.

Dean, Jodi. 2020. 'Neofeudalism: The End of Capitalism?', *Los Angeles Review of Books*, 12 May 2020. https://lareviewofbooks.org/article/neofeudalism-the-end-of-capitalism/.

Deleuze, Gilles and Guattari, Félix. 1983. *Anti-Oedipus: Capitalism and Schizophrenia.* Minneapolis, MN: Minnesota University Press.

Deleuze, Gilles and Guattari, Félix. 2004. *A Thousand Plateaus*. London: Continuum.

Dennett, Terry and Spence, Jo. 1979. *Photography/Politics: One*. London: Photography Workshop.

Derrida, Jacques. 2006. *Specters of Marx: The State of the Debt, the Work of Mourning and the New International*. London: Routledge.

Dewey-Hagborg, Heather. 2018. 'Generative Representation.' *unthinking photography*, December. https://unthinking.photography/articles/generative-representation.

Dewdney, Andrew. 2013. 'Curating the Photographic Image in Network Culture', in *The Photographic Image in Digital Culture*, ed. M. Lister, pp. 95–112. London: Routledge.

Dewdney, Andrew. 2019. 'The Networked Image: The Flight of Cultural Authority and the Multiple Times and Spaces of the Art Museum', in *The Routledge International Handbook of New Digital Practices in Galleries, Libraries, Archives, Museums and Heritage Sites*, ed. H. Lewi, W. Smith, and D. Vom Lehm, pp. 68–91. New York: Routledge.

Dewdney, Andrew, Dibosa, David and Walsh, Victoria. 2013. *Post Critical Museology: Theory and Practice in the Art Museum*. London: Routledge.

Dewdney, Andrew and Ride, Peter. 2013. *Digital Media Handbook, 2nd edn*. London: Routledge.

Drucker, Johanna. 2001. 'Digital Ontologies: The Ideality of Form in/and Code Storage – or – Can Graphesis Challenge Mathesis?' *Leonardo*, 34(2), pp. 141–145.

Dvorak, Tomas and Parikka, Jussi. 2021. *Photography Off the Scale: Technologies and Theories of the Mass Image*. Edinburgh: Edinburgh University Press.

Edwards, Steve. 2006. *Photography. A Very Short Introduction*. Oxford: Oxford University Press.

Emerling, Jae. 2012. *Photography: History and Theory*. New York: Routledge.

Ernst, Wolfgang. 2012. *Digital Memory and the Archive*. Minneapolis, MN: Minnesota University Press.

Fardy, Jonathan. 2012. 'Review: Jean Baudrillard, *Forget Foucault* [1976]. 2012 ISSN: 1832-5203', *Foucault Studies*, 13, pp. 184–187.

Farocki, Harun. 2001. '*Eye/Machine I*.' MACBA Collection. www.macba.cat/en/art-artists/artists/farocki-harun/eye-machine-i.

Fisher, Mark. 2009. *Capitalist Realism: Is There No Alternative?* Ropley: Zero Books.

Fisher, Mark. 2013. 'The Metaphysics of Crackle: Afrofuturism and Hauntology'. *Dancecult* 5(2), pp. 42–55.

Flusser, Vilém. 1983. *Towards a Philosophy of Photography*. London: Reaktion Books.

Flusser, Vilém. 2011 [1985]. *Into the Universe of Technical Images*. Minneapolis, MN: University of Minnesota Press.

Fontcuberta, Joan. 2015. *Le Mois de la Photo à Montréal: The Post-Photographic Condition.* Berlin: Kerber Verlag.

Foucault, Michel. 1990. *The Will to Knowledge; The History of Sexuality: 1,* trans. R. Hurley. London: Penguin Books.

Foucault, Michel. 2001. *The Order of Things.* London: Routledge.

Foucault, Michel. 2003. *The Birth of the Clinic: An Archaeology of Medical Perception.* London: Routledge.

Fraser, Andrea. 2005. '*From the Critique of Institutions to an Institutional Critique,' Artforum, September.* https://monoskop.org/images/b/b6/Fraser_Andrea_2005_From_the_Critique_of_Institutions_to_an_Institution_of_Critique.pdf.

Frosh, Paul. 2019. *The Poetics of Digital Media.* Cambridge: Polity.

Frizot, Michel. 1998. *A New History of Photography.* London: Konemann.

Fuller, Matthew. 2008. *Software Studies.* Cambridge, MA: MIT Press.

Gauntlett, David. 2000. *Web.Studies: Rewiring Media Studies for the Digital Age.* London: Hodder Education.

Genosko, Gary. 1999. *McLuhan and Baudrillard: Masters of Implosion.* London: Routledge.

Gibson-Graham, J.K. 2006. *A Postcapitalist Politics.* Minneapolis, MN: University of Minnesota Press.

Giroux, Henry A. 2009. *Zombie Politics and Culture in the Age of Casino Capitalism.* New York: Peter Lang.

Giroux, Henry A. 2010. 'Zombie Politics and Other Late Modern Monstrosities in the Age of Disposability,' *Policy Futures in Education,* 8(1). DOI: 10.2304/pfie.2010.8.1.1.

Gitelman, Lisa and Pingree, Geoffrey B. 2003. *New Media, 1740–1915.* Cambridge, MA: MIT Press.

Glazer, Ya'ara Gil. 2010. 'A New Kind of History? The Challenges of Contemporary Histories of Photography,' *The Journal of Art History,* 3, pp. 3–YGG/1.

Gomez-Cruz, Edgar and Lehmuskallio, Asko. 2016. *Digital Photography and Everyday Life: Empirical Studies in Material Visual Practices.* London: Routledge.

Hall, Stuart. 1995. *The Postcolonial Question.* London: Routledge.

Hall, Stuart and Massey, Doreen. 2010. 'Interpreting the Crisis,' *Soundings,* 44, p. 57.

Hand, Martin. 2012. *Ubiquitous Photography.* London: Polity.

Hansen, Mark B.N. 2015. *Feed-Forward: On the Future of Twenty-First-Century Media.* Chicago, IL: University of Chicago Press.

Haraway, Donna. 1985. 'A Cyborg Manifesto: Science, Technology, and Socialist-Feminism in the Late Twentieth Century,' in *Simians, Cyborgs and Women: The Reinvention of Nature.* London: Routledge.

Hartley, George. 2003. *The Abyss of Representation Marxism and the Postmodern Sublime.* Durham, NC: Duke University Press.

Hayles, Katherine. 1999. *How We Became Posthuman: Virtual Bodies in Cybernetics, Literature, and Informatics.* Chicago, IL: University of Chicago Press.

Haworth-Booth, Mark. 1997. *Photography: An Independent Art.* London: V&A Publishing.

Heidegger, Martin. 1977. *The Question Concerning Technology, and Other Essays.* New York: Harper and Row.

Heidegger, Martin. 2010 [1927]. *Being and Time: A Revised Edition of the Stambaugh Translation.* New York: State University of New York.

Henning, Michelle. 2018. *Photography: The Unfettered Image.* London: Routledge.

Hertz, Garnet D. and Parikka, Jussi. 2011. *Five Principles of Zombie Media.* www.digitalmanifesto.net/manifestos/23/.

Hertz, Garnet D. and Parikka, Jussi. 2012. 'Zombie Media: Circuit Bending Media Archaeology into an Art Method', *Leonardo*, 45(5). DOI: 10.1162/LEON_a_00438.

Hirsh, Robert. 2017. *Seizing the Light: A Social and Aesthetic History of Photography.* London: Routledge.

Hoelzl, Ingrid. 2014. *The Operative Image – An Approximation.* http://mediacommons.org/tne/pieces/operative-image-and-approximation.

Hoelzl, Ingrid and Marie, Remi. 2015. *Softimage: Towards a New Theory of the Digital Image.* Bristol: Intellect.

Holland, Patricia, Spence, Jo and Watney, Simon (eds). 1986. *Photography/Politics: Two.* London: Comedia.

Hooper-Greenhill, Eilean. 1992. *Museums and the Shaping of Knowledge.* London: Routledge.

Hunt, Tristram. 2018. 'Museums Must Confront the Big Issues', *The Art Newspaper*, 20 December. www.theartnewspaper.com/comment/museums-must-confront-the-big-issues.

Hughes, Joe. 2009. *Deleuze's Difference and Repetition.* London: Continuum.

Jenkins, Henry. 2006. *Convergence Culture.* New York: NYU Press.

Johansson, Troels Degn. 2002. *The New Image: On the Temporality of Photographic Representation after Digitalization.* Copenhagen: The IT University of Copenhagen.

Jones, Jonathan. 2003. 'I can't face it. I'm not going', *The Guardian*, 23 May.

Johnson, Richard. 2014 [1983]. 'Really Useful Knowledge: Radical Education and Working-Class Culture, 1790–1848', in *Education for Adults, Volume 2: Opportunities for Adult Education*, ed. M. Tight. London: Routledge.

Jurgenson, Nathan. 2019. *The Social Photo: On Photography and Social Media.* London: Verso.

Kember, Sarah. 1996. 'The Shadow of the Object: Photography and Realism,' *Textual Practice*, 10(1), pp. 145–163.

Kember, Sarah. 2008. 'The Virtual Life of Photography,' *Photographies*, 1(2), pp. 175–203.

Kember, Sarah and Zylinska, Joanna. 2012. *Life after New Media: Mediation as a Vital Process*. Cambridge, MA: MIT Press.

Kelner, Douglas. 2010. 'Reflections on Modernity and Post Modernity in Baudrillard and McLuhan,' in *Transforming McLuhan: Cultural, Critical, and Postmodern Perspectives*, ed. P. Grosswiler, pp. 179–202. New York: Peter Lang.

Kidron, Michael. 1989. *A Permanent Arms Economy*. London: SWP.

Kittler, Friedrich A. 2010. *Optical Media: Berlin Lectures 1999*, trans. A. Enns. Cambridge: Polity.

Klein, Naomi. 2008. *The Shock Doctrine: The Rise of Disaster Capitalism*. London: Penguin.

Kline, Kip. 2016. 'Jean Baudrillard and the Limits of Critical Media Literacy,' in *Educational Theory*, 66(5), pp. 641–656.

Knelman, Sara. 2015. 'Conflict, Time, Photography,' *Frieze*, 20 April. www.frieze.com/article/conflict-time-photography.

Kroes, Rob. 2007. '*Projecting National Identities through Cultural Diplomacy: The Case of The Family of Man*.' www.researchgate.net/publication/46698206_Projecting_National_Identities_Through_Cultural_Diplomacy_The_Case_of_The_Family_of_Man.

Kramp, Michael. 2012. 'Unburdening Life, or the Deleuzian Potential of Photography,' *Rhizomes, 23*. www.rhizomes.net/issue23/kramp/index.html.

Latour, Bruno. 1990. 'Drawing Things Together,' in *Representation in Scientific Practice*, ed. M. Lynch, and S. Woolgar, pp. 19–68. Cambridge, MA: MIT Press.

Latour, Bruno. 1993. *We Have Never Been Modern*. Cambridge, MA: Harvard University Press.

Latour, Bruno. 2004. 'Why Has Critique Run Out of Steam? From Matters of Fact to Matters of Concern,' *Critical Inquiry*, 30(2), pp. 225–248.

Latour, Bruno. 2007. *Reassembling the Social: An Introduction to Actor Network Theory*. Oxford: Oxford University Press.

Larsson, Lars Bang. 2010. 'Zombies of Immaterial Labor: The Modern Monster and the Death of New York,' *e-flux*, 15. www.e-flux.com/journal/15/61295/zombies-of-immaterial-labor-the-modern-monster-and-the-death-of-death/.

Laruelle, François. 2011. *The Concept of Non-Photography*. Cambridge, MA: MIT Press.

Lauro, Sarah Juliet. 2017. *Zombie Theory: A Reader*. Minneapolis, MN: University of Minnesota Press.

Lipovetsky, Gilles. 2002. *The Empire of Fashion: Dressing Modern Democracy*. Princeton, NJ: Princeton University Press.

Lipovetsky, Gilles. 2005. *Hypermodern Times*. Cambridge: Polity.

Lister, Martin. 2013 [1996]. *The Photographic Image in Digital Culture.* London: Routledge.

Lister, Martin. 2009 [2001, 2005]. 'Photography in the Age of Electronic Imaging' in *Photography: A Critical Introduction*, ed. L. Wells. London. Routledge.

Lister, Martin, Dovey, Jon, Giddings, Seth, Kelly, Keiran and Grant, Iain. 2009 [2003]. *New Media: A Critical Introduction.* London: Routledge.

Lovink, Geert. 2011. *Networks Without a Cause.* Cambridge: Polity.

Lütticken, Sven. 2013. 'World History and Earth Art.' *e-flux* #49. www.e-flux.com/journal/49/59999/world-history-and-earth-art/.

Lyotard, Jean-François. 1976. *The Condition of Postmodernity: A Report on Knowledge.* Manchester: Manchester University Press.

Lyotard, Jean-François. 1982. 'Presenting the Unpresentable: The Sublime,' *Artforum*, 20(8). www.artforum.com/print/198204/presenting-the-unpresentable-the-sublime-35606.

Lyotard, Jean-François. 1993. *The Inhuman: Reflections on Time*, trans. Bowlby, R. and Bennington, G. Cambridge: Polity.

Malevé, Nicolas. 2020. 'On the Data Set's Ruins.' *AI & Society*, https://doi.org/10.1007.

Manovich, Lev. 2001. *The Language of New Media*. Cambridge, MA: MIT Press.

Manovich, Lev. 2006 [1994]. 'The Paradoxes of Digital Photography,' in *The Photography Reader*, ed. L. Wells, pp. 240–249. London: Routledge.

Marcuse, Herbert. 2002. *Heideggerian Marxism*, ed. Richard Wolin and John Abromeit. Minneapolis, MN and London: University of Minnesota Press.

Marx, Karl. 2005. *The Eighteenth Brumaire of Louis Bonaparte*. New York: Mondail.

Marx, Karl and Engels, Friedrich. 1987. *The German Ideology: Introduction to a Critique of Political Economy.* London: Lawrence and Wishart.

Marx, Karl and Engels, Friedrich. 2015. *The Communist Manifesto*. London: Penguin.

Mason, Paul. 2015. *PostCapitalism: A Guide to Our Future.* Milton Keynes: Allen Lane.

May, John. 2019. *Signal, Image, Architecture: Everything is Already an Image.* New York: Columbia University Press.

Mbembe, Achille. 2019. *Necropolitics*. Durham, NC: Duke University Press.

McCocker, Anthony and Wilken, Rowen. 2020. *The Social Impact of the New Camera Consciousness.* London: Routledge.

McKenzie, Jai. 2014. *Light and Photomedia: A New History and Future of the Photographic Image.* London: Routledge.

McLuhan, Marshall. 2001 [1964]. *Understanding Media: The Extension of Man.* London: Routledge.

Mignolo, Walter. 2000. *Local Histories/Global Designs: Essays on the Coloniality of Power, Subaltern Knowledges and Border Thinking*. Princeton, NJ: Princeton University Press.

Mignolo, Walter. 2007. 'Delinking: The Rhetoric of Modernity, the Logic of Coloniality and the Grammar of Decoloniality,' *Cultural Studies*, 21(2–3), pp. 449–514.

Mirzoeff, Nicholas. 1999. *An Introduction to Visual Culture*. London: Routledge.

Mirzoeff, Nicholas. 2011. *The Right to Look: A Counterhistory of Visuality*. Durham, NC: Duke University Press.

Mißelbeck, Reinhold, Thielemann, Marianne Beiger and Goodrow, Gerald A. 2012. *20th Century Photography*. Berlin: Taschen.

Mitchell, William J. 1992. *The Reconfigured Eye: Visual Truth in the Post-Photographic Era*. Cambridge, MA: MIT Press.

Mitchell, William J. and Thomas, John. 1984. 'What is an Image?,' *New Literary History: A Journal of Theory and Interpretation*, 15(3), pp. 503–537.

Mitchell, W.J.T. 2005. *What Do Pictures Want?: The Lives and Loves of Images*. Chicago, IL: University of Chicago Press.

Moreman, Christopher M. and Rushton, Cory James. 2011. *Race, Oppression and the Zombie: Essays on Cross-Cultural Appropriations of the Caribbean Tradition*. London: McFarland.

Mulligan, Therese and Woolers, David (eds). 2005. *A History of Photography: From 1839 to the Present (The George Eastman House Collection)*. Berlin: Taschen.

Nechvatal, Joseph. 2009. *Towards an Immersive Intelligence: Essays on the Work of Art in the Age of Computer Technology and Virtual Reality 1993–2006*. New York: Edgewise Press.

Newhall, Beaumont. 1984 [1949]. *The History of Photography: From 1839 to the Present*. New York: Museum of Modern Art.

Nickel, Douglas. 2001. 'History of Photography: The State of Research,' *Art Bulletin*, 83(3), pp. 548–558.

O'Hagan, Sean. 2011. 'Britain's Photographic Revolution,' *The Guardian*, 30 October. www.theguardian.com/artanddesign/2011/oct/30/v-a-photographs-gallery-tate.

O'Hagan, Sean. 2014. 'The Scars of War: How Good is Photography at Capturing Conflict?,' *The Guardian*, 25 November. www.theguardian.com/artanddesign/2014/nov/25/conflict-time-photography-tate-modern-the-scars-of-war-sean-o-hagan.

O'Hagan, Sean. 2016. 'Sir Elton John: "I Collect for the Beauty, Not the Value. I'm in Awe of These Things",' *The Observer*, 13 November. www.theguardian.com/artanddesign/2016/nov/13/elton-john-interview-i-collect-photographs-for-the-beauty-tate-modern-radical-eye.

Osborne, Peter. 2013. *Anywhere Or Not At All*. London: Verso.

Osborne, Peter. 2010. 'Infinite Exchange: The Social Ontology of the Photographic Image,' *Philosophy of Photography*, 1(1), pp. 59–68.

Paglen, Trevor. 2014. 'Operational Images', *e-flux* #59. www.e-flux.com/journal/59/61130/operational-images/.

Parikka, Jussi. 2012. 'New Materialism as Media Theory: Medianatures and Dirty Matter', *Communication and Critical/Cultural Studies,* 9(1), pp. 95–100.

Paul, Christiane. 2003. *Digital Art*. London: Thames and Hudson.

Pickford, James. 2016. 'Elton John's Collection of Radical Photographs Go on Show at Tate Modern', *Financial Times,* 8 November. www.ft.com/content/c1c2162a-a5b7-11e6-8898-79a99e2a4de6.

Pink, Sarah. 2011. 'Sensory Digital Photography: Re-Thinking "Moving" and the Image', *Visual Studies,* 26(1), pp. 4–13.

Poole, Nick. 2010. *The Cost of Digitising Europe's Cultural Heritage*. A Report for the Comité des Sages of the European Commission. London: The Collections Trust.

Preziosi, Donald. 1991. *Rethinking Art History: Meditations on a Coy Science*. New Haven, CT: Yale University Press.

Price, Kati. 2018. '171 Clouds from the V&A Online Collection, 1630–1885 – a new commission by Penelope Umbrico', V&A Blog, 10 October. www.vam.ac.uk/blog/digital/171-clouds-from-the-va-online-collection-1630-1885-a-new-commission-by-penelope-umbrico.

Rancière, Jacques. 2004. *The Distribution of the Sensible*. London: Continuum.

Rancière, Jacques. 2007. *The Future of the Image,* trans. Gregory Elliott. New York: Verso.

Rancière, Jacques. 2009a. *The Emancipated Spectator,* trans. Gregory Elliott. New York: Verso.

Rancière, Jacques. 2009b. 'Notes of the Photographic Image', *Radical Philosophy,* 156. www.radicalphilosophy.com/article/notes-on-the-photographic-image.

Ribalta, Jorge, Wolf, Erika and Koltsov, Mikhail. 2011. *The Worker-Photography Movement 1926–1939*. Madrid: T.F. Editores.

Ricœur, Paul. 2004. *Memory, History, Forgetting*. Chicago, IL: University of Chicago Press.

Roberts, John. 2013. 'Solidarity and Form: Allan Sekula's Legacy', *Philosophy of Photography,* 4(1), pp. 139–141.

Roberts, John. 2016. 'Ideation and Photography: A Critique of François Laruelle's Concept of Abstraction', *Photographies,* 9(2), pp. 217–228.

Roberts, Bill. 2012. 'Production in View: Allan Sekula's *Fish Story* and the Thawing of Postmodernism.' Tate Papers No. 18. www.tate.org.uk/research/publications/tate-papers/18.

Robins, Kevin. 1996. 'Into the Image: Culture and Politics in the Field of Vision.' www.researchgate.net/publication/248817139_Into_the_Image_Culture_and_Politics_in_the_Field_of_Vision.

Rosenblum, Naomi. 2019. *A World History of Photography*. New York: Abeville Press.

Rosler, Martha. 2004. 'Post-Documentary, Post-Photography?', in *Decoys and Disruptions: Selected Writings, 1975–2001*, pp. 207–244, esp. 211. Cambridge, MA and London: MIT Press.

Røssaak, Eivind. 2011. 'Algorithmic Culture: Beyond the Photo/Film-Divide', in *Between Stillness and Motion: Film, Photography, Algorithms*, pp. 187–203. Amsterdam: Amsterdam University Press.

Rubinstein, Daniel. 2019. 'The New Paradigm', in *Fragmentation of the Photographic Image in the Digital Age,* ed. D. Rubinstein, pp. 1–8. London: Routledge.

Rubinstein, Daniel and Sluis, Katrina. 2013a. 'Algorithmic Photography and the Crisis of Representation', in *The Photographic Image in Digital Culture*, ed. M. Lister, pp. 22–40. New York: Routledge.

Rubinstein, Daniel and Sluis, Katrina. 2013b. 'Notes on the Margins of Metadata: Concerning the Undecidability of the Digital Image', *Photographies,* 6(1), pp. 151–158.

Said, Edward W. 2003 [1994]. *Orientalism*. London: Penguin.

Saucier, Nathan. 2017. 'Operational Images and The Interpretive Turn'. Master's thesis, Massachusetts Institute of Technology. https://cmsw.mit.edu/wp/wp-content/uploads/2018/02/371962369-Nathan-Saucier-Operational-Images-and-the-Interpretive-Turn.pdf.

Sealy, Mark. 2019. *Decolonising the Camera: Photography in Racial Time*. London: Lawrence and Wishart.

Sekula, Allan. 1978. 'Dismantling Modernism, Reinventing Documentary (Notes on the Politics of Representation', *The Massachusetts Review*, 19(4), pp. 858–883.

Sekula, Allan. 1981. 'The Traffic in Photographs', *Art Journal*, 41(1), pp. 15–25.

Sekula, Allan. 1986. 'The Body and the Archive'. https://monoskop.org/images/6/65/Sekula_1986_The_Body_and_the_Archive.pdf.

Sekula, Allan. 1995. *Fish Story*. Project and Exhibition, Centre for Contemporary Art, Rotterdam.

Sekula, Allan. 2017. *Photography Against the Grain: Essays and Photo Works, 1973–1983.* Halifax, NS: Press of Nova Scotia College of Art and Design.

Schuppli, Susan. 2020. *Media, Forensics, Evidence.* Cambridge, MA: MIT Press/Leonardo.

Sheehan, Tanya. 2014. *Photography, History, Difference*. Hanover, NH: Dartmouth College Press.

Shieber, Jonathan. 2019. 'More Investors are Betting on Virtual Influencers Like Lil Miquela', TechCrunch, 14 January. https://techcrunch.com/2019/01/14/more-investors-are-betting-on-virtual-influencers-like-lil-miquela/?sr_share=twitter&utm_source=tctwreshare&guccounter=1.

Sollitt, Lucy Rose. 2019. 'The Future of the Art Market', Creative United. https://fotamreport.creativeunited.org.uk/.

Star, Susan Leigh and Griesemer, Graham James R. 1989. 'Institutional Ecology, "Translations" and Boundary Objects: Amateurs and Professionals in Berkeley's Museum of Vertebrate Zoology, 1907–39,' *Social Studies of Science,* 19(3), pp. 387–420.

Steichen, Edward and Sandberg, Carl. 1996. *The Family of Man.* New York: Museum of Modern Art.

Steyerl, Hito. 2009. 'In Defence of the Poor Image,' *e-flux* #10. www.e-flux.com/journal/10/61362/in-defense-of-the-poor-image/.

Steyerl, Hito. 2010. 'The Politics of Art: Contempoary Art and the Transition to Post Democracy,' e-flux #21. www.e-flux.com/journal/21/67696/politics-of-art-contemporary-art-and-the-transition-to-post-democracy/.

Steyerl, Hito. 2012. *The Wretched of the Screen.* Berlin: Sternberg Press.

Steyerl, Hito. 2013. 'Too Much World: Is the Internet Dead?,' *e-flux* #49. www.e-flux.com/journal/49/60004/too-much-world-is-the-internet-dead/.

Steyerl, Hito. 2017. *Duty Free Art: Art in the Age of Planetary Civil War.* London: Verso.

Stiegler, Bernard. 1998. *Technics and Time: The Fault of Epimetheus.* Stanford, CA: Stanford University Press.

Stiegler, Bernard. 2009. *Technics and Time, 2: Disorientation.* Stanford, CA: Stanford University Press.

Stiegler, Bernard. 2016. *The Digital, Education, and Cosmopolitanism.* Berkeley, CA: University of California Press; adapted from a longer interview with Bernard Stiegler, published as 'Numérique, éducation, cosmopolitisme,' *Cités*, 63(3) pp. 13–36.

Stiegler, Bernard. 2018. *The Neganthropocene.* London: Open Humanities Press.

Stiegler, Bernard. 2019. *The Age of Disruption: Technology and Madness in Computational Capitalism.* Cambridge: Polity.

Srnicek, Nick. 2017. *Platform Capitalism.* London: Wiley.

Srnicek, Nick and Williams, Alex. 2015. *Inventing the Future: Postcapitalism and a World Without Work.* London: Verso.

Sontag, Susan. 2014 [1977]. *On Photography.* London: Penguin Classics.

Stallabrass, Julian. 2010. 'Museum Photography and Museum Prose,' *New Left Review,* 65. https://newleftreview.org/issues/ii65/articles/julian-stallabrass-museum-photography-and-museum-prose.

Strutt, Daniel. 2019. *Digital Image and Reality.* Amsterdam: Amsterdam University Press.

Solomon-Godeau, Abigail. 2017. *Photography After Photography.* Durham, NC: Duke University Press.

Suarez-Villa, Luis. 2012. *Globalization and Technocapitalism: The Political Economy of Corporate Power and Technological Domination.* London: Routledge.

Tagg, John. 1988. *The Burden of Representation: Essays on Photographies and Histories.* London: Palgrave Macmillan.

Tagg, John. 2009. 'Mindless Photography', in *Photography: Theoretical Snapshots,* ed. JJ. Long, A. Noble and E. Welch, pp. 16–30. London: Routledge.

Talbot, Fox. 1844. *The Pencil of Nature.* www.gutenberg.org/files/33447/33447-h/33447-h.html.

Thrift, Nigel. 2008. *Non-Representational Theory: Space, Politics, Affect.* Abingdon: Routledge.

Tight, Malcolm (ed.). 1983. '"Really Useful Knowledge": Radical Education and Working-class Culture, 1790–1848: Richard Johnson', in *Education for Adults*, pp. 18–37. London: Routledge.

Turkle, Sherry. 1997. *Life on the Screen: Identity in the Age of the Internet.* New York: Simon & Schuster.

Uricchio, William. 2011. 'The Algorithmic Turn: Photosynth, Augmented Reality and the Changing Implications of the Image', *Visual Studies*, 26(1), pp. 25–35.

Watney, Simon. 1980. 'Book Reviews: Photography/Politics: One, Photography Workshop, 1980', *Media, Culture and Society.* https://doi.org/10.1177/016344378000200309.

Warner Marien, Mary. 2014. *Photography, Fourth Edition: A Cultural History.* London: Laurence King.

Wells, Liz. 2015. *Photography: A Critical Introduction*, 5th edn. London: Routledge.

Wheeler, Michael. 2011. 'Martin Heidegger', Standford Encyclopedia of Philosophy. https://plato.stanford.edu/entries/heidegger/.

Winthrop-Young, Geoffrey. 2011. *Kittler and the Media.* Cambridge: Polity.

Williams, Raymond. 1961. *The Long Revolution.* London: Pelican.

Williams, Raymond. 2003 [1974]. *Television, Technology and Cultural Form.* London: Routledge Classics.

Wolin, Richard and Abomeit, John. 2002. *Herbert Marcuse: Heideggerian Marxism.* Minneapolis, MN: University of Minnesota Press.

Wrathall, Mark A. 2010. *Heidegger and Unconcealment: Truth, Language and History.* Berkeley, CA: University of California Press.

Zorzal, Bruno and Gabriel Menotti. 2017. 'Entrevista: Katrina Sluis, curadora digital da Photographer's Gallery, fala sobre fotografia e a cultura de imagens na internet'. *ZUM*, 13 November. https://revistazum.com.br/noticias/entrevista-katrina-sluis/.

Zylinska, Joanna. 2017. *Nonhuman Photography.* Cambridge, MA: MIT Press.

Zylinska, Joanna. 2021. 'Undigital Photography: Image-Making beyond Computation and AI', in *Photography Off the Scale*, ed. T. Dvorak and J. Parikka, pp 231–252. Edinburgh. Edinburgh University Press.

Notes

1 Forget Photography

1 Contemporaneity is used here precisely to indicate the disjuncture between the medium of photography and present time. It is also used with reference to the 'fiction' of the contemporary discussed by Peter Osborne in relationship to contemporary art (2013, p. 24).
2 Unthinking photography is the title of a blog started by the Digital Programme at the Photographers' Gallery, which overlaps with the discussion here regarding the new conditions of the computational image. It is also an oblique reference to the seminal photography theory title *Thinking Photography* (1982), edited by Victor Burgin.
3 Bruno Latour's analysis of the separation of human and non-human in modernist ontology is taken up at several points across the book and his account of the paradox of the modern is discussed in Chapter 6, when looking at the curatorial practices at Tate. Jean Baudrillard's reversal of signification to suggest that it is not reality which underwrites the sign, but the sign which guarantees the real, leads him on to suggest that reality is now a simulation. This idea is adopted to define the photographic image as a construction of the real.
4 Rancière defines the distribution of the sensible as, 'the system of self-evident facts of sense perception that simultaneously discloses the existence of something in common and the delimitations that define the respective parts and positions within it' (2004, p. 16).
5 Deleuze and Guattari use the terms deterritorialisation abstractly, as in leaving a plane of thought, a process always accompanied by reterritorialisation, as simultaneous process, rather than in anthropology as the material disconnection between place, dwelling and tradition and in the sociology of Anthony Giddens as a consequence of the global effects of migration. In photography deterritorialisation involves the uncoupling of the several material and cultural elements which maintain its unity and identity.
6 Reification. In the argument of the book, photography is a relational entity, whereas in common use it is turned into a fixed thing, with inherent attributes, as a consequence of what Marx defined as commodity fetishism in *Capital* Volume 1, chapter 1, section 4.
7 Knowledge considered as 'really useful' is an important emphasis in British cultural studies, particularly in the research undertaken at the Centre for Contemporary Cultural Studies at Birmingham University in the 1970s, informed by the work of Richard Hogget, Raymond Williams and Stuart Hall. It was expressed in Richard Johnson's chapter 'Really Useful Knowledge: Radical Education and Working-Class Culture, 1790–1848' in Malcolm Tight's *Education for Adults* (2014 [1983]).
8 Stuart Hall and Doreen Massey in conversation (2010).

2 Zombie Photography

1 Achille Mbembe developed the idea of necropolitics in order to understand how contemporary neoliberal global economic organisation produces forms of subjugation of life to the power of death, including the right of the state to impose social or civil death, the right to enslave others and of political violence. In this sense necrocapitalism creates precarious bodies located between life and death.

2 The importance of new writing on the decolonial falls upon method; see the writing of Walter Mignolo, particularly his 2007 paper, 'Delinking: The Rhetoric of Modernity, the Logic of Coloniality and the Grammar of Decoloniality,' *Cultural Studies*, 21(2–3), pp. 449–514. The question is not whether we should continue to decolonise all aspects of thought and practice, but how to go about it. The decolonial in the context of post-colonial cultures has to go beyond critique in developing new strategies for the production of knowledge and its educational exchange.

3 The project of seeing photography as a collective activity would unearth a great deal of research into numerous photography collectives, which if brought into a common frame of reference would begin to piece together an altogether different history of photography than that of the Western canon.

4 A good example of this tied relationship between dominant and oppositional photography can be found in *Photography/Politics: One* and *Two* (Dennett and Spence 1979; Spence, Holland and Watney 1986).

5 The idea of purification is used here as developed by Bruno Latour in his book *We Have Never Been Modern* to explain the constant dichotomies in separating out the world of objects and nature from the world of humans and culture. It is a concept used throughout the book and given prominence in Chapter 5, when considering modernism and photography.

6 Chrono-reflexivity is used here to emphasise that in post-industrial cultures the future no longer stretches to infinity, but has shrunk to the horizon of the present, in which time is experienced against itself, producing a multiplicity of times.

7 The unrepresentable is not simply that which was unrepresented, but that which couldn't be conceived within the apparatus of photography and is closer to Jean-François Lyotard's term 'unpresentable'; see his paper, 'Presenting the Unpresentable: The Sublime,' *Artforum*, April 1982, 20(8).

8 In Britain a new generation of artists and photographers, whose work explicitly expressed Black identity and heritage, emerged, including the work of Ingrid Pollard, David A. Bailey, Sunil Gupta, Maud Saulter, Zarina Bhimji and Lubaina Himid. Organisations included the picture agency Autograph, the Black Audio Film Collective and the Afircan and Asian Visual Artist's Archive, established by Eddie Chambers. It took two decades for their contributions to be fully recognised by national art organisations.

9 Alternative photography is an inadequate term to delineate a disparate array of photographic practices taken up by progressive, liberationist and oppositional movements, groups and individuals. The point being made is certainly not to point to the failure of radical practices, but rather that they operated within the media ecologies of their times.

3 Post-Photography

1 In *For a Critique of the Political Economy of the Sign* (1981, p. 161), Baudrillard argues that both constituent elements of the sign, the signifier and signified, are condemned to produce the arbitrariness of the sign, and hence, 'The salvation of use value from the system of exchange value, without realizing that use value is a satellite system in solidarity with that of exchange value: this is precisely the idealism and transcendental humanism of contents which we discover again in the attempt to rescue the signified from the terrorism of the signifier.' In effect, the real cannot be rescued by the system of signs.

2 Translation is a methodological approach developed in actor network theory (ANT) by Bruno Latour and Michel Callon to account for how networks come into view. For Latour translation is the complex process of negotiation in which meanings, claims and interests change and gain ground and through which power is always in play. Callon formalised the methodology of translation into four stages: problematisation; *Interessement*, corresponding to links between interests of actors; enrolment, the process of prioritisation; and mobilisation, which concerns the maintenance of a given network (Callon 1980, p. 211). Because translation processes involve human and non-human actors, it is also a methodological lens with which to look at hybrids.

3 The somewhat overwhelming academic focus upon the ontology of photography can be seen as a somewhat logical connection between material apparatus (technologies), the agency of objects and the phenomenology of affect. It is a strong strand of thought in the philosophy of photography, which is discussed in Chapter 4.

4 The post-human is recognised here as an important emphasis in the post-humanism and post-anthropocentric critical theory of Rosa Braidotti. Here Braidotti emphasises the need to rethink subjectivity as a collective assemblage that encompasses human and non-human actors, technological mediation, animals, plants, and the planet as a whole.

4 Philosophy, Technology and Photography

1 For Heidegger, unconcealment is a form of truth, which he developed from the ancient Greek philosophical term *aletheia*, meaning not concealed. Heidegger developed the idea beyond that of revealing what was concealed to mean both bringing things to awareness and creating the context within which things can be what they are. Unconcealment was a key method in Heidegger's ontological philosophy. See Mark A. Wrathall's *Heidegger and Unconcealment: Truth, Language and History* (Berkeley, CA: University of California, 2010).

2 Herbert Marcuse attempted to reconcile Marxism with Heidegger's theoretical notion of existence. See Herbert Marcuse's *Heideggerian Marxism*, ed. Richard Wolin and John Abromeit (Minneapolis, MN and London: University of Minnesota Press, 2002).

5 Photography and Modernism: A Case Study of Tate Modern and Tate Britain

1 'Tate and Photography,' Tate Modern press release, 25 November 2014. www.tate.org.uk/press/press-releases/tate-and-photography.

2 The work of Richard Hamilton in the Tate collection is of note in this respect.

3 Tate made links with the Centre for the Study of the Legacies of British Slave-ownership at the University of Central London. www.ucl.ac.uk/lbs/.

4 Jones (2003).

5 Simon Baker, quoted in O'Hagan (2011).

6 Pierre Bourdieu's grand work *Distinction: A Social Critique of the Judgement of Taste* was published in 1979 as a result of five years of empirical fieldwork, which he analysed in terms of French social classes, developing the thesis that non-financial social assets, such as education, which promote upward social mobility lead to a form of cultural capital and a form of social distinction through the exercise of taste.

7 The term independent photography gained currency in the UK in the 1970s and 1980s to refer to forms of documentary and community photography which lay outside of mainstream media and art photography.
8 This is a reference to Peter Osborne's discussion of the contemporary as a 'fiction', given the multiple sites of modernity, in his book *Anywhere Or Not At All* (2013).
9 Bishop (2012b).
10 *Cybernetic Serendipity*, curated by Jasia Reichardt for the ICA in London in 1968, focused upon art using computer-generated sound and imagery, including work by Peter Zinovieff, Gustav Metzger, Gordon Pask, Bruce Lacey and Nam June Paik amongst others.
11 'Modelling Cultural Value within New Media Cultures and Networked Participation' (2014, ed. V. Walsh, A. Dewdney and E. Pringle) was the official report of a collaborative project between Tate, The Royal College of Art and London South Bank University, funded by the Arts and Humanities Research Council. https://openresearch.lsbu.ac.uk/download/66a9dd85351993da36f5cd7064ac07567e3779ba1284f864af6c702d7f0c41fa/5286113/Cultural%20Value%20and%20Networked%20Participation%20copy.pdf.
12 The paradoxical present is a term developed by Gilles Lipovetsky in *Hypermodern Times* (2005) and discussed at several points throughout this volume. It is used to designate a period beyond the postmodern in which time's horizon has shrunk to that of the present and in which time is pitted against time.
13 Pickford (2016).
14 'Room Guide: Objects, Perspectives, Abstractions', Tate. www.tate.org.uk/whats-on/tate-modern/exhibition/radical-eye-modernist-photography-sir-elton-john-collection-7.
15 O'Hagan (2016).
16 Knelman (2015).
17 O'Hagan (2014).
18 Cumming (2014).

6 Photography and Heritage: A Case Study of the Victoria and Albert Museum

1 The European Commission Directorate General for Communications Networks, Content and Technology has conducted extensive policy, coordination as well as funding actions to supplement Member States' cultural policy in areas of digitisation and online access to cultural material and digital preservation. Funding for digitisation projects has been made available through the EU Horizon 20:20 scheme.
2 Tony Bennet's 'Exhibitionary Complex' (1988), Stuart Hall's 1995 essay *The Postcolonial Question* and Edward Said's *Orientalism* (1994) laid the foundations for the postcolonial critique of the museum.
3 Tristram Hunt, widely quoted, here in the Museums and Heritage Advisor, 12 October 2016. https://advisor.museumsandheritage.com/news/va-photography-centre-launches-major-exhibition-daguerreotype-digital/.
4 *Tripe | Ruff* by Thomas Ruff. www.vam.ac.uk/articles/tripe-ruff.
5 Price (2018).
6 Poole (2010).
7 about.artsandculture.google.com.
8 Caines (2013).

7 The Image after Photography

1 Gilles Lipovetsky (2002) argues that fashion embodies the modern preoccupation with the body and the cult of appearance, which he analyses as historically linked to the rise of democratic values. With the advent of the computationally rendered body the image is seamlessly fused with the fashion image.

2 www.thecut.com/2018/05/lil-miquela-digital-avatar-instagram-influencer.html.

3 Online fashion marketing is globally orientated and keenly aware of the appeal of ethnic diversity in developed centres of migration as well as the growing affluence of sectors of industrialising populations.

4 In January 2019, TechCrunch reported that Lil Miquela's creators had closed a $125-million investment round led by Spark Capital. Suddenly, virtual influencers were the future of ads. The future of fashion. The future of commerce. See Shieber (2019).

5 The information here is gleaned from the website of the Smartphone Photographer. https://thesmartphonephotographer.com/.

6 Hitachi, Ltd. developed Japan's first lensless camera in 2016. It is expected that, by incorporating the lensless camera onto a variety of devices. Major camera companies such as Nikon and Canon are using built in algorithms for HDR and Deep Learning AF.

7 Røssaak (2011).

8 Uricchio (2011).

9 Nicolas Malevé undertook a PhD, based at the Centre for the Study of the Networked Image at London South Bank University, in collaboration with the Photographers' Gallery, London.

10 Although not specifically discussed, Donna Haraway's early political insights into technology used as an anti-patriarchal force, opening up a new nature and subjectivity, in her seminal 1985 work 'A Cyborg Manifesto: Science, Technology, and Socialist-Feminism in the Late Twentieth Century' remain prescient in contemporary debate.

8 The Politics of the Image

1 I am aware that this is a classical Marxist formulation, which could be dismissed as reductive and determinist in the contemporary cultural context, perpetrating a grand narrative. As Latour points out, 'What is to be done, then, with such sleek, filled-in surfaces, with such absolute totalities? Turn them inside out all at once, of course; subvert them, revolutionize them – such was the strategy of those modernists par excellence, the Marxists. Oh, what a lovely paradox!' (1993, p. 126). Nevertheless, even if we abandon ideas of system for networks, capitalism is a grand narrative, even if intellectually hollowed out. The mode of production and the mode of reproduction are retained as performing an abstract unity, of what might otherwise be defined as related assemblages.

2 In *Reassembling the Social* (2007), Latour gives an account of actor network theory, but with the caution that there are three things wrong with the account: the words actor, network and theory. The burden of the book is not only to give an account of an alternative social semiotic method of analysis, but to show how sociology as a discipline had objectified the object it set out to study.

3 In a post-human, non-dualist model, the 'system of capitalism' would be reconstituted as the 'matrix of power' (Braidotti 2017) based upon bio-power and thus avoiding the exteriorisation of capitalist ontology. Here capitalism is retained as the overriding term which defines a total of relations.

4 The Defence Advanced Research Projects Agency (DARPA) is an R&D agency of the United States Department of Defense responsible for the development of emerging technologies for use by the military, founded in 1958, and supporting research on decentralised command structures, of which the Internet is the prime example.

5 The photomontages of John Heartfield directed against the Nazi alliance of capitalism and militarism in Germany are examples of photography used to express conceptual ideas about the nature of power.

6 Ribalta et al. (2011).

7 *Photography/Politics: One* (1979) was edited by Terry Dennett, David Evans, Silvia Gohl and Jo Spence (Photography Workshop with Comedia). *Photography/Politics: Two* (1986) was edited by Patricia Holland, Simon Watney and Jo Spence (Routledge). Contributor Simon Watney summed up the purpose of the two volumes in saying that as socialists it is crucial to call into question institutional practices and to do everything possible to engage in the widest possible debate about politics of visual imagery.

8 A number of not-for-profit community arts organisations developed in UK urban centres, including in London's Blackfrairs Settlement, Paddington Printshop, Centreprise and Camerawork, as well as Bootle Arts in Liverpool, Clocktower in Birmingham and Lighthouse in Brighton.

9 www.moma.org/collection/works/152791.

10 'Allen Sekula [sic]: Aerospace Folktales and Other Stories,' Columbus Museum of Art. www.columbusmuseum.org/blog/news_room/allen-sekula-aerospace-folktales-and-other-stories/.

11 Docklands Community Poster Project development supported Lorraine Leeson and Peter Dunn at East London University; the Cockpit Gallery touring exhibitions were archived at London South Bank University, with other community photographic activists teaching photography in universities, such as Antony Luvera at Coventry.

12 Banerjee (2008).

13 Dean (2020).

14 Suarez-Villa (2012).

15 Beller (2016).

16 *Workers Leaving the Googleplex*, 12 mins, 2011.

17 Sollitt (2019).

18 Copies are owned by Fondazione Sandretto Re Rebaudengo in Turin, Fondation Louis Vuitton pour la Création in Paris, De Pont Museum of Contemporary Art in Tilburg, Ellipse Foundation in Cascais, Art Institute Chicago, a private collection in Greece and the Museum of Modern Art in New York.

19 Steyerl (2012, chapter 6, p. 92).

20 Lucy Rose Sollitt's 'The Future of the Art Market.' Published in partnership with Arteïa, Arts Council England, Creative Scotland and DACS, digital-only publication released on 12 November 2019. https://www.creativeunited.org.uk/services/the-future-of-the-art-market/

21 The Centre for the Study of the Networked Image was established in 2012 and is supported by the School of Arts and Creative Industries at London South Bank University. It is co-directed by Geoff Cox, Annet Dekker, Andrew Dewdney and Katrina Sluis and runs a postgraduate research programme in collaboration with international cultural organisations.

22 Brett Rogers quoted in O'Hagan (2011).

23 These guidelines have been produced for the Arts Council by MTM (Richard Ellis and Ed Corn) and the Space (Fiona Morris, John White and Owen Hopkin). MTM is a research and strategy consultancy specialising in arts and culture, media and technology. MTM's clients include Google, the BBC and the National Gallery. MTM ran the three-year Digital Culture survey for Arts Council England and Nesta. The Space is a community interest company founded by Arts Council England and the BBC. It brings arts organisations, artists and technology together to make great art and reach new audiences. It commissions digital projects, builds digital skills and helps arts organisations reach more people using digital media.

24 The Esmée Fairbairn Foundation is a registered charity founded in England in 1961. It is one of the largest independent grantmaking foundations based in the UK, funding organisations which aim to improve the quality of life for people and communities in the UK. Its grant to the Photographers' Gallery for the Digital Programme provided the salary for the digital curator. https://esmeefairbairn.org.uk.

25 https://thephotographersgallery.org.uk/media-wall.

9 The Hybrid Image

1 Stiegler (2016).

2 Transmediale was founded in 1988 as VideoFilmFest, a part of the Berlin annual International Forum of New Cinema, shifting its focus over time to a dialogue with television and multimedia. Transmediale is now part of a network of partners focused on art and digital culture.

3 The Serpentine's Research and Development Platform is a project developed by Ben Vickers and Kay Watson. One of its projects is to model R&D for arts and technology innovation. www.serpentinegalleries.org/arts-technologies/rd-platform/.

4 The position of post-criticality is taken up in considering the positions of Latour and Baudrillard further in this discussion. It was also elaborated as a research practice position in Dewdney, Dibosa and Walsh (2013).

5 'Nicholas Mirzoeff, visual activist, in conversation with Sonya Dyer: Illuminating Black Histories,' YouTube, National Portrait Gallery, 26 November 2020. www.youtube.com/watch?v=aMxdMjqW_Cc&ab_channel=NationalPortraitGallery. This interview was shown as part of the Understanding British Portraits professional network live event 'From De-colonial to Anti-colonial: What's Next for Museum Interpretation?' on 26 November 2020.

6 Steyerl (2013).

7 Steyerl (2009).

Index

Note: Page numbers in *italic* refer to Figures